ADVANCES IN COMMUNICATIONS AND MEDIA RESEARCH

VOLUME 4

Advances in Communications and Media Research

Advances in Communications and Media Research Volume 4
William H. Miller (Editor)
2007. ISBN 1-60021-189-5

Advances in Communications and Media Research Volume 3
Patula E. Grossmark (Editor)
2007. ISBN 1-60021-188-7

Advances in Communications and Media Research Volume 2
Anthony V. Stavros (Editor)
2003. ISBN 1-59033-613-5

Advances in Communications and Media Research Volume 1
Anthony V. Stavros (Editor)
2002. ISBN 1-59033-215-6

ADVANCES IN COMMUNICATIONS AND MEDIA RESEARCH

VOLUME 4

WILLIAM H. MILLER
EDITOR

Nova Science Publishers, Inc.
New York

For permission to use material from this book please contact us:
Telephone 631-231-7269; Fax 631-231-8175
Web Site: http://www.novapublishers.com

Library of Congress Cataloging-in-Publication Data
Available upon request.

ISBN 13 978-1-60021-189-8
ISBN 10 1-60021-189-5

Published by Nova Science Publishers, Inc. ✢New York

CONTENTS

PREFACE

In a society predicated on information, the media has a pervasive presence. From government policy to leisure television, the information age touches us all. The papers collected in this book constitute some of today's leading analyses of the information industry. Together, these essays represent a needed foundation for understanding the present state and future development of the mass media. Current trends in communications as well as media impact on public opinion are studied and reported on.

In chapter 1, just as journalism has had to adapt to the Internet, so has media law, and those changes will have an impact on the development of online media. Changes in copyright law are affecting public discourse – by criminalizing new forms of conduct, reducing access to certain types of information, altering patterns of distribution and changing assumptions behind fair use. Changes in defamation law are redefining the qualifications for journalistic protection. Internet data collection is putting new demands on privacy law, while raising ethical questions for media sites juggling opportunities for profitability and their users' rights. And the borderless quality of the Internet is presenting new challenges for personal jurisdiction in legal cases related to these issues and others handled domestically and abroad.

Weekly newspapers enter the millennium increasing in number and circulation. This is because entry barriers are lower in weekly newspaper markets than in daily newspaper markets. As discussed in chapter 2, many weeklies also distribute some or all of their circulation free. Advertisers unwilling to pay prices charged by large dailies may find weeklies attractive. Weeklies may also offer content that strengthens readers' ties to other members of geographic, religious or social communities. The number of weeklies owned by groups is increasing, which may allow those papers to extend scale economies in the production of advertising and news. Newspaper companies that own dailies are acquiring weeklies to extend their reach into new segments of their markets. Weeklies appear well positioned to thrive in the new millennium.

Ageism, or discrimination and prejudice on the basis of age, promotes a climate that promotes negative or positive attitudes and beliefs based on chronology. The literature on this subject is vast with much of the focus on negative attitudes and behavior toward older adults. In fact, ageism may be negative or positive and aimed at younger and older people. Chapter 3 discusses the literature on ageism as a background for the evaluation instrument known as the "*Nuessel-Stewart Ageism Rating Instrument.*" This assessment vehicle is a form with "yes/no" questions, a numerical scale, Charles Osgood's semantic differential format, and a section for written commentary. The purpose of this instrument is to provide instructors with a format for assessing negative and positive verbal and visual characterizations of ageism across the chronological spectrum. While it is an instrument to assess ageism, it also serves as a pedagogical tool for classroom use and for homework assignments through practical, applied assignments. The purpose of this presentation is four-fold: (1) to discuss the research on ageism; (2) to provide an example of the "*Nuessel-Stewart Ageism Assessment Instrument*;" (3) to exemplify its usage with text and image alone or in combination; and (4) to provide a useful bibliography on this topic.

As American newspaper readership continues to decline, critics foresee adverse effects for the present-day public-affairs news product. Declining competition hints that newspapers will seek short-term profits over quality. But chapter 4 offers that newspapers have a strategic choice because of a more "dynamic" view of competition as posited by Bridges, Litman and Bridges (2002). The discussion uses the underlying assumptions inherent in the concept of competition and suggests that news and the newsroom are abstract, momentary phenomena that shift and move as fluidly as consumer behavior, prices, competition and management. In short, news is a multifaceted marvel and is discussed as such via the lenses of news production, the framing of competition, the economics of technology, the dynamics of costs, and the nuances of timing.

Two prominent television events in the past two years placed increased attention on the FCC and its indecency regulations. The airing of the 2003 Golden Globe Awards and the subsequent ruling by the FCC's Enforcement Bureau, coupled with the controversy surrounding the 2004 Super Bowl halftime show, brought broadcast indecency to the forefront of the congressional agenda during the 108th Congress. Several bills were introduced, but not enacted, in the 108th Congress to increase the penalties imposed for broadcast indecency and to prohibit the broadcast of certain words and phrases in any grammatical form. Similar legislation aimed at increasing penalties has been introduced in the 109th Congress (H.R. 310, S. 193, S. 616). Other legislation addressing specific types of programming has also been introduced (H.R. 1420 and H.R. 1440). Chapter 5 provides background on the two events in question, discusses the legal evolution of the FCC's indecency regulations, and provides an overview of how the current regulations have been applied and pending legislation. The final section of the report considers whether prohibiting the broadcast of "indecent" words regardless of context would violate the First Amendment.

Competing for access to the conversational floor is sometimes thought of as a way of 'accomplishing power' in interaction. However, research on 'interactional power' easily becomes ideological to the degree where analysts risk losing sight of the very phenomena they claim to investigate. A symptom of this within studies of discourse organization is the relatively free traffic between concepts such as overlap, turn competition and interruption. Even more deceptive is the acclaimed – yet seldom thoroughly examined – relation of these conversational phenomena to dominance and power, both of which are often treated as discursive consequences of participants' gender or social status (e.g., Bargiella-Chiappini & Harris, 1996; Berger, 1994; Ferguson, 1977; Huls, 2000; Palmer, 1989; West & Zimmerman, 1983). In chapter 6, the authors will leave the latter problems aside, focusing mainly on issues of overlapping speech and interruption. Specifically, drawing upon conversation analytic work on interruptive procedures (e.g., Bilmes, 1997; Hutchby, 1993), they will present a couple of instances of interruption occurring in disparate multiparty settings. Specifically, the authors will argue that the work of 'being interrupted' – a defining feature of interruptions in CA studies – may be accomplished by other parties than the interruptee, and also that participants may engage in such work to achieve local interactional goals. They begin this chapter with a brief presentation of some issues of conversational organization relevant for our understanding of interruption in talk.

As an extended shamanising media event, the intention of which was to promote reconciliation and healing as a primary process of nation building, the core of the media TRC can be seen as pedagogic, providing a new frame and purpose for the nation and its institutions. As in all public events (Handelman 1990) this desired interpretation was contained in the very sequencing and ordering of the event, in the form of the three committees which both generated a moral process and enacted a model of healing for individual and nation. This moral process was constructed within at times contested discourses of truth and reconciliation, with imagining and promoting a necessary unity underpinning the whole. In chapter 7, the outcome of the whole process constructed as a media event was a narrative of the nation which, given its preoccupation with the past can also be understood as an alternative history, a new, inclusive representation of the past which could serve the interests of nation building.

In: Advances in Communications and Media Research, Volume 4 ISBN 1-60021-189-5
Editor: William H. Miller, pp.1-18

Chapter 1

MEDIA LAW AND THE INTERNET

Ashley Packard*
University of Houston-Clear Lake, Houston, TX, USA

ABSTRACT

Just as journalism has had to adapt to the Internet, so has media law, and those changes will have an impact on the development of online media. Changes in copyright law are affecting public discourse – by criminalizing new forms of conduct, reducing access to certain types of information, altering patterns of distribution and changing assumptions behind fair use. Changes in defamation law are redefining the qualifications for journalistic protection. Internet data collection is putting new demands on privacy law, while raising ethical questions for media sites juggling opportunities for profitability and their users' rights. And the borderless quality of the Internet is presenting new challenges for personal jurisdiction in legal cases related to these issues and others handled domestically and abroad.

INTRODUCTION

The Internet is still a relatively small part of the revenue stream for most media companies, but it is the only media segment experiencing growth and attracting young people to news (Project for Excellence, 2004). Media companies that recognize the Internet's emerging significance as a news source are developing Web sites to take advantage of the new medium's capabilities. And while the Internet has not required a wholesale revision in the basic principles of journalism, it has required journalism to adapt. Media law is also adapting to meet the needs of the Internet, and although the law cannot keep up with the Internet's speed, it is having a powerful effect on its development. This chapter will discuss

* Correspondence to: Dr. Ashley Packard, email: packard@uhcl.edu

changes in law and new legal issues related to copyright, defamation, privacy and jurisdiction that are likely to have an impact on the development of online media.

COPYRIGHT, CONGRESS AND THE COURTS

No area of law has been more affected by the Internet than copyright. Likewise, in recent years, no area of media law has received more publicity. Due to the recent spate of subpoenas for online music theft issued by the Recording Industry Association of America, this once dry area of law has generated thousands of news articles. Internet users, who rarely thought about copyright, are now recognizing the law's potential to affect them directly.

But recent changes in copyright law are also affecting public discourse in other ways people may not realize – by criminalizing new forms of conduct, reducing access to certain types of information, altering patterns of distribution and changing assumptions behind fair use.

Copyright law provides creators of expressive works with a monopoly right to control the use of their works for a limited period of time to "promote the progress of science and the arts" (*U.S. Constitution*, Art 1, §8, cl.8). The underlying assumption of the law is that people will only produce new works that contribute to society if their intellectual efforts are protected.

As our culture and economy have become increasingly dependent upon the exchange of digital information that is easy to reproduce and distribute, copyright infringement has become a significant problem. Congress has responded by making copyright law more restrictive.

Congress first addressed the vulnerability of copyrighted material online when it passed the No Electronic Theft Act (1997). The NET Act closed a loophole in copyright law that discouraged prosecution of nonprofit file trading. Congress crafted the legislation in response to *United States v. LaMacchia* (1994), a case against an electronic bulletin board operator who encouraged his users to upload and download computer games, but never profited directly. Because copyright law did not address willful infringement in which no one profited, the government charged LaMacchia with wire fraud. A federal court in Massachusetts dismissed the claim and advised Congress to change copyright law if it wanted to criminalize such behavior.

The Net Act changed the law's definition of financial gain to include "receipt or expectation of receipt of *anything* of value, including…copyrighted works" [emphasis added] (17 USC § 101). It also clarified threshold limits for criminal copyright infringement.[1]

[1] Any person who in a 180-day period willfully infringes a copyright either for financial gain or by the reproduction or distribution of one or more copyrighted works with a value exceeding $1,000, can be imprisoned for up to one year or fined up to $100,000. Infringement involving the exchange of 10 or more works valued at $2,500 in the same period constitutes a felony punishable by up to $250,000 in fines or imprisonment of 5 years.

Copyright Extension

One year later, Congress passed the Sonny Bono Copyright Term Extension Act (1998), which extended the term of copyright protection by 20 years. A copyrighted work is now protected for the life of the creator plus 75 years, or in the case of corporate works, for 95 years. Critics of the extension – which was both prospective and retrospective – challenged its constitutionality (*Eldred v. Ashcroft*, 2003). They argued that copyright protection, which has been extended 11 times in the last 40 years, is a limited monopoly right, not perpetual ownership "on the installment plan" (Hearings on S. 4839, 1995). Nevertheless, the Supreme Court upheld the Copyright Term Extension Act as a reasonable way of aligning U.S. law with the more restrictive protections of the European Union for future trade purposes.

DCMA

In addition to lengthening term limits, Congress moved to protect digital works in other ways with the Digital Millennium Copyright Act of 1998. Internet users who deem information a public resource have never been particularly respectful of intellectual property laws, so copyright holders who are serious about protecting their works have learned to use technology, such as digital watermarks and encryption, to their advantage. The Digital Millennium Copyright Act supports those efforts by criminalizing the circumvention of a "technological measure that effectively controls access" to a work protected under federal copyright law (17 U.S.C. § 1201).

This provision has been controversial because of its negative impact on speech and press. The law has been used to censor publishers, journalists and scientists from sharing articles and research about encryption. Some censorship has been official. In *Universal City Studios v. Corley* (2001), the Second Circuit Court of Appeals upheld an injunction barring an online magazine from publishing, or even linking to, a software code that could be used for infringing purposes. But fear of the law has also resulted in self-censorship. Publishers have dropped books about computer security flaws, fearing they might be construed as a "circumvention device" [2] (Unintended Consequences, 2004; Schiesel, 2003). Reporters have shied away from encrypted documents that might have been valuable to a story[3] (McCullagh, 2002). Scientists have been pressured into withdrawing research that described encryption flaws in software[4] (Unintended Consequences, 2004; RIAA, 2001).

[2] John Wiley & Sonsin 2003 dropped a book on a Microsoft security flaw that the publisher had commissioned from security researcher Andrew Huang due to fear of liability under the DMCA.

[3] Declan McCullagh wrote about his decision not to read publicly available documents about airline safety on the Transportation Security Administration Web site because they were encrypted. Although a source had given him the password to break the code, he was too afraid of DMCA liability to use it.

[4] In 2001, a Princeton scientist and fellow researchers responded to a public challenge issued by a multi-industry group called Secure Digital Music Initiative to defeat an electronic watermark used to protect digital music. When the scientists succeeded and tried to present their research at an academic conference, the same group threatened them with legal action. Also in 2001, Russian programmer Dmitry Sklyarov was jailed for several weeks in the United States after speaking at the DEFCON conference in Las Vegas. He was charged with working on software that could convert Adobe's e-Book format into Portable Document Format (PDF) files, a technology with legitimate uses. The Department of Justice ultimately permitted Sklyarov to return home, but prosecuted his employer, ElcomSoft, under DMCA criminal provisions. A jury acquitted the company in 2002.

In crafting the DMCA, Congress intended to impede piracy, not legitimate uses of information. But the legislation is written very broadly – so broadly, in fact, that it seems to have inadvertently created a new right – the right to control access to a work. Copyright law entitles copyright holders to control reproduction, distribution, adaptation, public display and public performance of their copyrighted works (17 U.S.C. § 106). But it has never entitled a copyright holder to prevent others from reading, seeing or listening to a work. In fact, the copyright statute acknowledges that there are times when the use of copyrighted works, even without the copyright holder's permission, is fair (17 U.S.C. § 107). Although the statute does not supply the circumstances when fair use applies, it does mention that favored uses include news reporting, criticism, comment, teaching, scholarship and research.

The Digital Millennium Copyright Act, however, prevents the use of anti-encryption software to gain access to copyrighted material, regardless of its intended use. In other words, even if the use of the material would be lawful, circumventing the copyright holder's encryption measures to gain access to it is not. As the use of encryption grows, this statute will make it more difficult to gain access to material that relates to legitimate reporting.

One aspect of the Digital Millennium Copyright Act that has drawn praise from First Amendment advocates is the law's safe harbor provision for Internet Service Providers. Congress crafted a provision in the Act that shields ISPs from copyright liability for subscribers' acts, as long as the ISPs have no knowledge of them (17 U.S.C. § 512). Once notified of the infringement, though, ISPs must disable access to the material to retain their immunity. This section of the law also provides that copyright holders may subpoena the names of infringers from Internet service providers. In fact, the Recording Industry Association of America is using this provision to force ISPs to provide the names of subscribers caught downloading MP3 files.

Freelancers

In general, courts seem to be more protective of copyright as well. Publishers suffered a copyright blow that affected their ability to distribute some articles electronically when the Supreme Court sided with freelancers against them in *New York_Times v. Tasini* (2001). Jonathan Tasini, president of the National Writers Union, sued *The New York Times* and several other media organizations, challenging the publishers' right to reuse freelance articles in digital databases without their writers' permission. The publishers argued that inclusion of freelance articles in their databases was simply a revision of their original collective works, which did not require the writers' permission. The Court disagreed, comparing inclusion of the articles in a database to a second publication.

The freelancers' victory was short-lived. Rather than gaining more profit from additional uses of their works, Tasini and his fellow petitioners found that publishers simply stiffened their contracts to play hardball against them. *The New York Times* and other newspapers began requiring freelancers to sign over future rights to digital publication and, in some cases, all rights to the work (Nelson, 2001). Freelancers who refused were dropped from the publications' pool of writers. In fact, some newspapers, conditioned future work on the

---

Blackboard, an educational software company, also used a DMCA threat to stop the presentation of research on security vulnerabilities in its products at the InterzOne II conference in Atlanta.

demand that freelancers transfer electronic publication rights retroactively to past print articles. A Massachusetts court found that the ultimatum was not an unfair or deceptive business practice (*Marx v. Globe Newspaper Co.*, 2002)·

Individually, freelancers have no power to demand payment for additional uses of their work, and a Catch 22 in antitrust law prevents them from banning together for collective bargaining (Johnston, 2002). Conventional labor unions are allowed to bargain collectively because they enjoy a statutory exemption from antitrust laws that does not apply to independent contractors like freelance writers.

Postings and Hyperlinks

Newspapers, like freelance writers, want to protect their copyrighted work from unauthorized use and have challenged Web sites that post their full-text articles without authorization. *The Los Angeles Times* and *The Washington Post* successfully sued the owner of the Free Republic Web site, which allowed users to post full-text newspaper articles, for contributory copyright infringement (*LA Times v. Free Republic*, 2000). Users of the bulletin board site posted articles along with their comments about the pieces. Subsequently, other users added their comments to extend the "discussion." Despite public discourse associated with the postings, the court determined that use of the articles was not fair and fined the New Republic $1 million in statutory damages. The decision, although limited to California, is significant because it is now a common practice among organizations engaged in public relations to post news articles about their activities on their Web sites.

Newspapers have also tried to protect their content against unauthorized hyperlinks and framing, practices that are inherent to the Web but at times cause confusion about the source of information. But because most of these cases have been settled out of court, there is still very little case law on the subject.

Simple hyperlinks that transfer a user from one Web site to another without creating a false impression about the source of the content are not a violation of copyright law (*Ticketmaster Corp. v. Tickets.com*, 2000). But the issue of deep linking, the practice of linking to Web pages buried in an organization's site that may not contain identifying information or advertising, is yet to be resolved. A Scottish case involving one newspaper's deep links to another newspaper's articles was settled out of court before the issue could be decided (*Shetland Times Ltd. v. Wills*, 1997). A Holland court rejected the argument that deep linking is illegal. The court upheld the right of a news headline site to link directly to articles (Cramb, 2000). But, in the United States, the Ninth Circuit recently reversed a decision that found deep linking to be a fair use. In *Kelly v. Arriba Soft Corporation* (2003), a photographer sued a search engine operator, Arriba, that reproduced thumbnail copies of his work without his authorization and used inline links to his Web site, creating the impression that larger versions of his photos resided on Arriba's site. A federal district court granted summary judgment to Arriba, finding both uses fair. The Ninth Circuit agreed that the search engine's use of the thumbnails was fair, noting that the use was transformative, [5] but thought

[5] The term "transformative" has purposely been left open to interpretation to give courts flexibility in fair use cases, but in Cambell v. Acuff-Rose, the U.S. Supreme Court described a transformative work as one that "adds

the use of inline links to the larger images was unfair (*Kelly v. Arriba*, 2002). The appeals court later withdrew its decision and replaced it with another explaining that the district court's ruling on inline linking was premature because neither party had moved for summary judgment on the issue (*Kelly v. Arriba*, 2003).

Courts have had mixed reactions to sites containing links to material that is potentially infringing or an invasion of privacy. A federal district court in Utah preliminarily enjoined owners of a Web site critical of Mormon Church doctrine from knowingly linking to other Web sites that posted parts of the Church's copyrighted handbook (*Intellectual Reserve v. Utah Lighthouse Ministry*, 1999). The Second Circuit Court banned *2600* magazine from linking to any site that publishes DeCSS software, which descrambles DVD content (*Universal City Studies v. Corley*, 2001). A Norwegian programmer had developed the software to allow people to play DVDs on computers running the Linux operating system. When *2600* magazine published a story about the software that included a link to it, eight movie studios sued. During coverage of the controversy, mainstream newspapers, including *The New York Times* and the *Village Voice*, which also operate within the Second Circuit's jurisdiction, linked to the DeCSS site (Electronic Frontier Foundation, 2004). Presumably the decision would also have an impact on their future coverage of the issue. In contrast, the California Supreme Court held that publishing information about decoding of DVDs merits strong First Amendment protection. The Supreme Court ordered the Court of Appeals to reevaluate evidence in a trademark suit filed against Andrew Bunner for republishing DeCSS on his website. A Florida judge was also roundly criticized on First Amendment grounds for issuing a temporary injunction to stop a man from posting stories about his relationship with a former Miss Vermont or linking to her Web site (Golab, 2003). Because he used the site to endorse a book he had written on "pick-up lines," the plaintiff sued him for appropriation of her name for commercial purposes. Miss Vermont dropped the suit when the defendant appealed to a federal court.

FAIR USE

Fair use is the primary legal defense against copyright infringement and a particularly important mechanism used to prevent copyright law from impinging upon First Amendment rights. But, in an era of increasing accessibility to copyrighted works, courts are recognizing fair use on fewer occasions (Bunker, 2002; Zimmerman, 1998). Several legal commentators have observed that courts seem to be looking for transformative uses of copyrighted material, and if none exist courts are less likely to acknowledge fair use. In assessing a fair use claim, a court will consider the purpose, nature and amount of the use, and the effect of the use on the market for the work (17 USC § 107). Transformation, which relates to the purpose and character of the use, is not a requirement to succeed in a fair use defense. But following *Campbell v. Acuff Rose* (1994), a parody case in which the Supreme Court emphasized transformation heavily in its analysis, many courts are of the opinion that if users are not generating something new from the original work, the use is unfair.

something new, with a further purpose or different character, altering the first with new expression, meaning, or message" (510 U.S. 569, 579 (1994).

Transformation can be a tricky concept, because a secondary user must transform the work enough to fall within fair use, without violating the right of the copyright holder to control derivative uses of a work. The ease of cut and paste makes it tempting to incorporate others' work into news articles, photo montages and graphic illustrations. Publishers need to be extremely vigilant in training employees to avoid copyright infringement, which is so easy to accomplish on the Internet. As recent plagiarism scandals related to Jayson Blair of *The New York Times* and Jack Kelley of *USA Today* have demonstrated, no newspaper is immune from this threat (Barry, Barstow, Glater, Liptak and Steinberg, 2003; Johnson, 2004).

DEFAMATION AND THE INTERNET

Although copyright is the one area of law most changed by the Internet, defamation is the area of Internet law most litigated (Juman and Greenwald, 2002).

In some ways, defamation online is no different than defamation offline. It still emanates from false communications that damage a plaintiff's reputation. It still requires plaintiffs to demonstrate the same six elements: publication of the allegedly defamatory material, identification of the plaintiff, the libelous nature of the material, the falsity of the statement or impression left, fault on the part of the defendant, and the actual harm that resulted.

But in other ways the online environment and the technology used with it seem to make a difference in the way defamation should be treated, and the courts and Congress have struggled to sort out these issues.

Defining Journalists

One question often asked in relation to defamation law now that so many Web sites provide information is "Who qualifies as a journalist?" Journalists in the United States are granted "extra breathing room" in cases of defamation, a policy intended to encourage aggressive news coverage. In fact, the Supreme Court has acknowledged that "debate on public issues ...may well include vehement, caustic, and sometimes unpleasantly sharp attacks..." (*New York Times v. Sullivan*, 1964)

One such sharp attack came from self-appointed cyber journalist Matt Drudge who accused former presidential aide Sidney Blumenthal of spousal abuse in his online column, The Drudge Report. Although Drudge retracted the charge the next day, Blumenthal sued for defamation. A federal judge considering the case ruled that Drudge, who had no formal training or experience in journalism before starting The Drudge Report, was not a "reporter, news gatherer, or journalist" (*Blumenthal v. Drudge*, 1998, p.57, n. 18). Although traditional journalists have never embraced Drudge, they grudgingly admit that he has altered the rules of online journalism. He scooped other media, when he was the first to report President Clinton's affair with Monica Lewinsky, and claims that his Web site logs 1 billion hits a year (Glassman, 1998). Today, he would probably earn heightened protection under the First Amendment.

Courts have since determined that Web sites disseminating news and information are "entitled to all the First Amendment protections accorded a newspaper/magazine or journalist in defamation suits" when they publish regularly about matters of public concern (*Banco Nacional de Mexico v. Rodriguez* , 2001, Media Defendants, 2). It is not yet clear how Web logs – Web sites that function as online journals – fit into this equation. There are 3 million Web logs, or blogs, on the Internet, ranging in content and personality (Blacker, 2003). Some are personal journals. Some are lists of links to particular subjects. Others report regularly on topics of public interest. In fact, both the Democratic and Republican National Committees agreed to issue press credentials to some bloggers for their 2004 conventions (Jesdanun, 2004, Weiss, 2004). A recent MIT study found that despite their belief that they are liable for what they publish, bloggers generally do not believe that people could sue them for what they write on their blogs (Viegas, 2004).

Statutes of Limitations

Another issue to be resolved online is how to assess the statute of limitations for online media. In the case of a printed publication, thousands of issues may be disseminated, but together they constitute a single publication resulting in one cause of action that must be filed before the relevant statute of limitations expires. Depending on the state in which the case is tried, the statute of limitations for defamation ranges from one to three years. But in the case of online publication, when defamatory material is continuously available for viewing, the issue is less clear-cut.

The New York Court of Appeals recently addressed this issue to determine the appropriate statute of limitations in an Internet defamation suit (*Firth v. State*, 2002). The question was whether the statute of limitations begins when the material is initially published online and ends one year later, or whether it should be renewed each time the material is accessed or each time the Web site is changed in some way and, technically, reposted. Considering the impact the decision would have on free speech, the court determined that the statute of limitations begins when the material is first put online, and that modification of another part of the Web site does not constitute a republication triggering a renewal of the limitation period. The court reasoned that if the implication of posting new information to a site meant that the potential for a defamation suit was renewed, Web masters might restrict the amount of new information they disseminate.

Courts have not always been so successful in sorting out differences between on- and offline law, particularly when the issue was liability for publication of libelous material. Publishers are always liable for the defamatory content of their publications, under the assumption that they exercise editorial control over content. In contrast, distributors of those same publications are shielded from liability under the theory that they cannot possibly be expected to read every publication they sell. A policy forcing them to do so would inevitably result in the reduction of publications distributed, reducing access to information.

Publishers or Distributors

Courts looking at the first online cases struggled with whether to characterize online service providers as publishers or distributors for purposes of liability. One court held that an online service provider that offered subscribers access to what it described as a "library of material" was a distributor (*Cubby v. Compuserve*, 1991). Another court held that an online service provider that offered subscribers access to electronic bulletin boards that were subject to editorial control could be characterized as a publisher (*Stratton Oakmont v. Prodigy*, 1995). The answer to escaping liability for subscriber posts seemed to lie in avoiding editorial review altogether and functioning more like a common carrier, such as a telephone company does.

Congress, which was looking for ways to clean up the Internet, had to step in to ensure that ISPs continued to monitor content. To encourage a policy of review, Congress crafted an exemption from liability for subscriber defamation in the Communications Decency Act of 1996. Section 230 of the CDA stipulates that "no provider or user of an interactive computer service shall be treated as the publisher or speaker of any information provided by another information content provider" (47 U.S.C. § 230(c)(1)). Although the Supreme Court struck down parts of the Communications Decency Act that unconstitutionally restrained speech, the provision granting immunity to Internet Service Providers remains intact.

The open question related to the provision is whether a Web site that permits third parties to post content, but is not actually an Internet service provider, should qualify as an "interactive computer service" and enjoy the same immunity. At least one state court has found this to be the case. In *Schneider v. Amazon.com, Inc.* (2001), an author sued Amazon.com because a third party had posted an allegedly defamatory book review on Amazon.com's Web site. The Washington Court of Appeals dismissed the action on the grounds that Amazon.com is immune to suit under the CDA. The Ninth Circuit also held that listserv owners are covered by the exemption, as long as the person who provided the information intended it for publication on the Internet (*Batzel v. Smith*, 2003). Decisions such as these will become increasingly important to media Web site operators as they offer more interactive services to attract subscribers.

Privacy and Data Collection

Privacy law seems to be the area struggling the most to keep up with the Internet. The ease of information acquisition and exchange in the digital age is unparalleled. Meanwhile traditional privacy protections offer very little control over personal information that is collected and exchanged online. The media are certainly benefiting from collections of information in databases, particularly as journalism relies more on computer-assisted reporting. From a business perspective, data gathered from online audiences can translate into additional sources of revenue. But privacy also introduces potential hazards for the media, most notably an increased risk of compromising audience trust and the ability to operate effectively in countries with more rigorous privacy protections.

Public Records

A variety of data once only available in musty basement files are now stored in digital databases often accessible through the Internet. Public records, such as birth certificates, marriage licenses, property valuations, divorce proceedings, bankruptcy filings, Social Security benefits and wills, as well as civil and criminal court proceedings, have traditionally been invaluable reporting tools. They are even more valuable in a digital format that enables journalists to do computer-assisted reporting that would have been impossible with paper records. Take *The St. Louis Post-Dispatch*, for example. When the newspaper's staff cross-referenced electronic voter registration and death records, they discovered the dead were still voting in East St. Louis, Illinois (Novak and Landau, 1990).

But media organizations are also gathering digital information from their readers for business purposes that are rarely publicized. Media Web sites are among the worst offenders when it comes to surreptitiously collecting data without disclosing their privacy policies. In fact, only a little more than a third of the media companies are currently posting their privacy policies, although two thirds of them are tracking their users' movements online or collecting their personal information (Pryor and Grabowicz, 2003).

Data gathered from subscribers can help media Web sites learn more about what subscribers want to read and the types of services that attract them. Demographic data collected online also can be useful to attract advertisers to Web sites, which are still struggling to make a profit (Project for Excellence, 2004). But media, which already struggle with public perception about journalistic ethics, should also be aware that surreptitious data collection is not particularly popular among the public, who are concerned about the use of their personal information. A Pew Internet Project Report found that 84 percent of Internet users are concerned about businesses or people they don't know getting personal information about themselves or their families (2002).

Privacy law in the United States has evolved to respond to technological challenges, but in a piecemeal fashion, rather than as a response to a unified policy. Consequently, privacy protection for information collection over the Internet is somewhat patchy.

Privacy Policies

The Constitution implicitly protects some aspects of privacy, particularly through the Fourth Amendment, which prohibits unlawful search and seizure by government entities. But it does not protect information collected for legitimate business purposes (*Smith v. Maryland*, 1979). Nor does it protect personal information voluntarily provided to third parties "even if the information is revealed on the assumption that it will be used only for a limited purpose and the confidence placed in the third party will not be betrayed" (*United States v. Miller*, 1976, p. 443). In fact, courts have used the Constitution's First Amendment and commerce clause to invalidate state legislation protecting Internet privacy (*ACLU of Georgia v. Miller*, 1997; *State v. Heckel*, 2001; see also Fenno, 2001).

Certain federal statutes protect aspects of personal privacy. The most comprehensive of these is the Privacy Act of 1974, which prohibits the federal government from collecting and disseminating certain types of information. But the statute is inapplicable to private entities that collect personal data (5 U.S.C. § 552a). Congress has considered comprehensive protection against data collection from children. The Children's Online Privacy Protection Act of 1998 prohibits commercial Web sites from collecting personal information from children under 13 without parental permission (15 USC § 6501 *et. seq*). Although there is no comprehensive legislation to guard the privacy of information collected from adults, Congress has been moving in that direction. The Federal Trade Commission Act, which prohibits unfair or deceptive practices in the marketplace, enforces the promises commercial Web sites make to consumers in their privacy policies about how personal information is collected, used, shared and secured (15 U.S.C. § 45). Unfortunately the choice of whether or not to enact a policy is entirely voluntary. Also, the Gramm-Leach-Bliley Financial Modernization Act of 1999 entitles customers of financial institutions to prevent the institutions from providing their personal financial information to e-mail marketers or online profilers (15 U.S.C. § 6801, *et seq.*; 12 C.F.R. § 40).

Other federal statutes protect bits and pieces of privacy. The Electronic Communications Privacy Act (18 USC § 2511), for example, protects against the interception and disclosure of any wire, oral or electronic communication, which covers telephone and e-mail communications, but not information collected on Web sites through registration forms or cookies that gather personal information and record click streams. The Telecommunications Act of 1996 regulates the use of personal information by telecommunications carriers (47 USC § 222). Initially, the FCC interpreted the law to mean that carriers had to obtain permission from their customers to share personal information for marketing purposes. But the Tenth Circuit struck down that interpretation in *U.S. West, Inc. v. FCC* (1999) as an unconstitutional restriction on commercial speech. The court said:

> Although we may feel uncomfortable knowing that our personal information is circulating in the world, we live in an open society where information may usually pass freely. A general level of discomfort from knowing that people can readily access information about us does not necessarily rise to the level of a substantial state interest....(*U.S. West, Inc. v. FCC*, 1999, p. 1234-35)

State laws provide some statutory protection for privacy. But these vary by jurisdiction and are subject to weak enforcement. Actions against data collectors generally come under common law invasion of privacy claims against intrusion upon seclusion, public disclosure of private facts, portrayal in false light and misappropriation of identity for commercial purposes (Groemminger, 2003, p. 837).

Of these four theories, the two most likely to apply to data collection on the Internet are intrusion upon seclusion and public disclosure of private facts, but neither fully meets the requisite requirements. In consideration of a privacy claim, courts generally seek to determine whether there is a reasonable expectation of privacy that another party has violated without justification, whether the violation is highly offensive to the reasonable person, and whether harm resulted from that violation (DeVries, 2003, p. 307). Courts have not considered data collection to be highly offensive, reasoning that individuals could foresee that a Web site might collect data (Groemminger, 2003, p. 837).

Moreover, individuals voluntarily provide much of the personal data collected in bits and pieces that only become intrusive after they are aggregated into a larger profile. Once an individual gives up control of this information, it is considered public and, under the First Amendment, distribution of public information cannot be limited without a compelling state interest (DeVries, 2003, p. 308).

EU Regulations

In response to increasing concern about privacy among its member states, the European Union has enacted a privacy policy that comprehensively protects EU citizens from unauthorized data collection. Under the Data Protection Directive, which took effect in 1998, 25 member states of the European Union must require information collectors to get individuals' consent before collecting and using their personal information or transferring it to a third party (Directive 95/46/EC of the European Parliament, 1995). Also, information collectors are required to disclose the purpose for the data collection, as well as to allow individuals to view data collected from them and correct that data if necessary.

The EU directive makes it clear that member states will not engage in electronic commerce with any country that does not protect its citizen's privacy rights (Groemminger, 2003, p. 841). The United States has responded to EU concerns by requiring the Department of Commerce to develop a database of companies in the United States that adhere to privacy standards similar to those mandated by the EU (U.S. Department of Commerce Safe Harbor, 2004). For the time being, EU concerns seem to have been assuaged, but the voluntary arrangement may not last. At some point, to facilitate electronic commerce between the United States and European Union, Congress may be compelled to adopt the EU directive on privacy standards, as it adopted the EU's extended term for copyright protection in 1998.

PERSONAL JURISDICTION AND THE INTERNET

Exacerbating the already complex legal issues related to the Internet is the fact that it has no physical boundaries, potentially leaving content providers vulnerable to legal action in a number of jurisdictions. Courts may assert specific jurisdiction over a defendant if the defendant has sufficient "minimum contacts" with a forum state to satisfy traditional notions of fair play (*International Shoe Co. v. Washington*, 1945, p. 316). In *Calder v. Jones* (1984), the Supreme Court held that personal jurisdiction is proper when a defendant intentionally commits a tortuous act aimed at the forum state, knowing that the brunt of the harm will be felt in that state. The *Calder* "effects test" focuses on where the effects of the tort are felt most. Federal circuits have applied the test to determine Internet jurisdiction with varying results.

State Jurisdiction for Businesses

Relying on *Calder*, the Ninth Circuit Court of Appeals concluded that Washington state could exercise personal jurisdiction over a Colorado company that "purposefully injected itself" into the Washington market by offering information of interest to Washington residents on its Web site (*Northwest Healthcare Alliance v. Healthgrades.com, Inc.*, 2002, p. 341). The Ninth Circuit's ruling allowed Washington-based Northwest Healthcare Alliance to sue Healthgrades.com, a Web-based company that rates health care services nationwide, for an "average" ranking it considered defamatory.

In contrast, the Fourth Circuit Court of Appeals held that Virginia could not exert jurisdiction over two Connecticut newspapers that published allegedly defamatory stories about a Virginia prison warden on the Internet (*Young v. New Haven Advocate*, 2002). *The New Haven Advocate* and the *Hartford Courant* published articles on their Web sites that discussed conditions in a Virginia prison that housed Connecticut prisoners. The Court of Appeals held that minimum contacts were not established through publication on the Web because the newspapers were not aimed at a Virginia audience. To come to this conclusion, the court considered a number of factors: neither paper solicited subscriptions from Virginia residents; neither had offices or employees in Virginia, and neither derived substantial revenue from goods or services rendered in Virginia. Likewise, the individual defendants were not Virginia residents and had no assets or business relationships there. But, from a journalistic perspective, one factor considered by the court was disturbing. The Court of Appeals seemed to focus on the fact that reporting for the story had been rather shallow. In its opinion, the court points out that.

No one from either newspaper, not even the reporters, traveled to Virginia to work on the articles about Connecticut's prisoner transfer policy. The two reporters...made a few telephone calls into Virginia to gather some information for the articles. Both interviewed by telephone a spokesman for the Virginia Department of Corrections. All other interviews were done with people located in Connecticut. (*Young v. New Haven Advocate*, 2002, p. 260).

The court's analysis seems to imply that a more thorough investigation could have triggered personal jurisdiction.

In an earlier copyright infringement action, the Fourth Circuit had already concluded that "a person's act of placing information on the Internet" is not sufficient to "subject that person to personal jurisdiction in each state in which the information is accessed" (*ALS Scan v. Digital Service Consultants*, 2002, p. 712). Otherwise, the court reasoned, "the person placing information on the Internet would be subject to personal jurisdiction in every state."

To determine Internet jurisdiction, the Fourth Circuit, like many other courts, has also relied on the so-called *Zippo* test, which now seems to be falling out of favor (*ALS* at 713). The sliding scale test developed by a Pennsylvania court in *Zippo Manufacturing Co. v. Zippo Dot Com Inc.* (1997), considers personal jurisdiction to be "directly proportionate to the nature and quality of commercial activity that an entity conducts over the Internet" (*Zippo* at 1124). In other words, it looks at the Web site's level of interactivity. Sites actively engaged in Internet commerce fall at one end of the continuum, while those passively engaged in information distribution fall at the other. The weakness of the test lies in its application to the middle ground, where courts are forced to make subjective decisions about Web sites that are interactive, but not fully engaged in e-commerce, as well as in the nature of the test itself, which focuses on a Web site's level of interactivity rather than the degree to which it purposely targeted the forum state (Dawson, 2004).

International Jurisdiction

The Internet also crosses national boundaries, raising questions about international jurisdiction. This situation is of particular concern to U.S. media, which enjoy broader protection for expression than do media in other countries. Disparity of press protection is obvious in more restrictive regimes, such as China, where religious expression is prohibited, but press freedoms are also more limited in traditionally liberal countries like Australia and France.

Only days before *Young v. New Haven Advocate* was decided, the High Court of Australia claimed jurisdiction over a U.S. publisher in *Dow Jones v. Gutnick* (2002). The case involves a suit over an allegedly defamatory news story on the Dow Jones *Barron's* Web site. The Australian court reasoned that "defamation is to be located at the place where the damage to reputation occurs…." (*Dow Jones*, 2002, 44)· It also determined that Australian law should apply in the case and that it should be tried in a Victoria court. In 2003, the reporter who wrote the story for Dow Jones took the unusual step of appealing the decision to the United Nations Human Rights Committee, alleging Australia is violating his right to free speech (McLachlan, 2003). The committee has no power to overturn domestic law; it can only render an opinion on the issue.

This is not the first time a foreign court has asserted jurisdiction over an American company for its Internet activities. An earlier case concerned freedom of expression. A French organization, the International League Against Racism and Anti-Semitism, sued Yahoo!, claiming its auction site violated French law, which prohibits the purchase, possession or exhibition of Nazi memorabilia. After determining that the Yahoo! site was a "threat to the internal public order," a Paris court ordered Yahoo! to block French citizens' access to anti-Semitic material (Dawson, 2004, p. 641). Yahoo complied with the order on its French site site. But the Internet company asked a California federal district court for a declaratory judgment barring enforcement of the order in the United States (*Yahoo! v. La Ligue Contre Le Racism Et L'Antisemitisme*, 2001). The court granted the judgment, holding that an order that "prohibits the sale or display of items based on their association with a particular political organization and bans the display of Web sites based on the authors' viewpoint" would amount to prior restraint (*Yahoo!*, 2001). In other words, upholding the French order in the United States would violate the company's First Amendment rights.

The European Union is grappling with the issue of international jurisdiction for civil torts. Its member states are considering a proposed law, known as Rome II, which would require a court in which a civil suit is filed for defamation or invasion of privacy to follow the law of the country where the damage is sustained. The law would provide an exception when the laws of the plaintiff's country are considered incompatible with the "fundamental principles" of freedom of expression (Commission of the European Communities, 2003). The United Kingdom's European Union Committee of the House of Lords is against the proposal. Citing complications likely to arise when Internet publishers and service providers are subjected to the laws of multiple countries, the committee has proposed an alternative "country of origin" policy (Rojas, 2004). Such a policy would require that any suit for defamation brought against an online publication be filed in the country where editorial control is exercised. This would align protections for online media with those of broadcast media that already operate under a country of origin rule under other EU Directives (House of Lords, 2004).

International jurisdiction is a wild card for media publishing on the Internet. Case law related to it is still in its infancy, and countries are in conflict over choice of law. Because national interpretations of the meaning and importance of press freedom are inextricably tied to the culture and history of a particular country, we are unlikely to come to a shared standard of freedom of the press that will apply in civil cases. In the short term, companies that operate on the Internet will likely use technology to block certain countries from viewing specific content. But, with an interest in protecting the Internet's growth, if not for freedom of the expression then for preservation of global commerce, in the long term the international community will have to come to terms with this issue.

Conclusion

The online world holds great promise for media growth, but to prosper there media owners will have to navigate through changing laws related to intellectual property, defamation, and privacy as well as the great divide emerging among international jurisdictions. Those who carefully forge ahead will write the story of the future of media.

References

ACLU of Georgia v. Miller, 977 F. Supp. 1228 (N.D. Ga. 1997) (finding a Georgia statute precluding anonymous use of the Internet violated the *First Amendment*).

ALS Scan v. Digital Service Consultants, Inc., 293 F.3d 707, 712 (4th Cir.2002).

Banco Nacional de Mexico v. Rodriguez, Index No. 603429/00 (N.Y. App. Div. 2001), unpublished opinion. Retrieved May 11, 2004 from http://www.eff.org/Cases /BNM_v_Narco_News/20011205_decision.html

Barry, D., Barstow, D. Glater, J., Liptak A., and Steinberg J. (2003, May 11). Times Reporter Who Resigned Leaves Long Trail of Deception. *The New York Times*, p. 1.

Batzel v. Smith, 351 F. 3d 904, 1034 (2003).

Blacker T. (2003, June 10). On Planet Blog, Everyone Can be a Celebrity. *The Independent (London)*, p. 11.

Blumenthal v. Drudge and America Online, Inc., 992 F. Supp. 44, 57 n.18 (D.D.C. 1998).

Bunker, M. D. (2002). Eroding Fair Use: The "Transformative" Use Doctrine After Campbell. *Communication Law and Policy*, *7*, 1-24.

Calder v. Jones, 465 U.S. 783, 788-90 (1984).

Campbell v. Acuff-Rose, 510 U.S. 569, 579 (1994).

Children's Online Privacy Protection Act of 1998, Pub. L. No. 105-277, 112 Stat 2681-728 (1998) (codified at 15 U.S.C. § 6501 (2000)).

Commission of the European Communities, *Regulation of the European Parliament and the Council on the Law Applicable to Non-Contractual Obligations ("Rome II")* Brussels, 22.07.2003 Com (2003) 427 final 2003/0168 (COD). Retrieved Oct. 12, 2004 from Europa.eu.int/eur-lex/en/ com/pdf/2003/com2003_0427en01.pdf

Communications Decency Act of 1996, Pub. L. No. 104-104, tit. V, 110 Stat. 56, 133-43 (1996)(codified at 47 U.S.C. § 230(c)(1)).

Copyright Act of 1976, Pub. L. No. 94-553, 90 Stat. 2541 (codified at 17 U.S.C. 101-803 (2000)).

Cramb, G. (2000, Aug. 23). Dutch Papers Fail in Cyber Space. *Financial Times,* Retrieved May 7, 2004 from http://news.ft.com/

Cubby v. CompuServe, Inc., 776 F. Supp. 135, 140-41 (S.D.N.Y. 1991).

Dawson, C. (2004). Creating Borders on the Internet: Free Speech, the United States, and International Jurisdiction. *Virginia Journal of International Law, 44*, 637-653.

Digital Millennium Copyright Act of 1988, Pub. L. No. 105-304, 112 Stat. 2860 (codified at 17 U.S.C. § 512, 1201).

Directive 95/46/EC of the European Parliament and of the Council of 24 October 1995, Eur. O.J. L281/31 (Nov. 23, 1995). Dow Jones and Company Inc. v Gutnick 2000 HCA 56 (10 December 2002). Retrieved March 19, 2004 from http://www.austlii.edu.au /au/cases/cth/high_ct/2002/56.html

Eldred v. Ashcroft, 538 U.S. 916 (2003).

Electronic Communications Privacy Act of 1986, Pub. L. No. 99-508, 100 Stat. 1848 (2000) (codified as amended at 18 U.S.C. 2510-2710 (2000)).

Electronic Frontier Foundation (July 3, 2002). Media Release: *2600 Magazine* Won't Seek Supreme Court Review in DVD Case. Retrieved May 7, 2004, from http://www.eff.org/IP/Video/MPAA_DVD_cases/20020703_eff_2600_pr.html

Federal Trade Commission Act, 15 U.S.C. § 45 (2003).

Fenno, E. (Jan.-Feb. 2001). Federal Internet Privacy Law. *South Carolina Law, 12*, 36-41.

Firth v. State, 98 N.Y.2d 365; 775 N.E.2d 463 (2002), affg 287 A.D.2d 771 (2001).

Glassman, J. (1998, June 9). Matt Drudge, E-Journalist. *Washington Post*, p. A15.

Golab, A. (2003, May 8). Two-time Miss Vermont Says Her Name Exploited. *Chicago Sun-Times*, p. 4.

Gramm-Leach-Bliley Financial Modernization Act, Pub. L. No. 106-102 § 222, 113 Stat. 1388 (codified at 15 U.S.C. 6801-27 (2000)).

Groemminger, B (2003). Note: Personal Privacy on the Internet: Should It be a Cyberspace Entitlement? *Indiana Law Review, 36*. 827-855.

Hearings on S. 4839, The Copyright Term Extension Act of 1995, Before the Senate Judiciary Comm., 104th Cong., 1st Sess. (1995). Available at 1995 WL 1052-4355 (testimony of Professor Peter Jaszi, Washington College of Law, American University).

House of Lords Committee on European Union, United Kingdom Parliament. (2004, March 30) *The Rome II Regulation report with evidence 8th report of session 2003-04.* Retrieved Oct. 12, 2004, from http://www.publications.parliament.uk/pa/ld200304/ldselect/ldeucom/66/6602.htm#evidence

Intellectual Reserve, Inc. v. Utah Lighthouse Ministry, Inc., 75 F. Supp. 2d (1999).

International Shoe Co. v. Washington, 326 U.S. 310, 316 (1945).

Jesdanun, A. (2004, July 28). GOP Invites Bloggers to Convention, National Political News, *Associated Press Online.* Retrieved Oct. 12, 2004 from Lexis Nexis Academic database.

Johnson, P. (2004, Jan. 14). Similar wording is found in '*Post*,' *USA TODAY* stories. *USA Today*, p. 3A.

Johnston, J. (2002, June 3). Free-Lance Writers Lost in Cyberspace; Will Congress remedy a quirk in antitrust law that prevents free-lancers from organizing and pressing their cause? *Legal Times*, p. 28.

Juman R. and Greenwald, M. (2002, February 4). Beware of Liability for Web Postings. *New York Law Journal*, p. S8.

Kelly v. Arriba Soft Corp, 280 F.3d 934, 948 (2002)(opinion withdrawn and replaced with Kelly v. Arriba Soft Corp., 336 F.3d (9th Cir. 2003)).

Kelly v. Arriba Soft Corp. 336 F.3d 811, 822 (9th Cir. 2003) (opinion filed Feb. 6, 2002, withdrawn July 7, 2003, refilled July 7, 2003).

L.A. Times v. Free Republic, Case No. CV 98-7840, 2000 U.S. Dist. LEXIS 5669, 54 U.S.P.Q.2D (BNA) 1453, 1469-71 (C.D. Cal. 2000); punishment phase in *L.A. Times* v. Free Republic, Case No. CV 98-7840, 2000 U.S. Dist. LEXIS 20484, 56 U.S.P.Q.2D (BNA) 1862 (C.D. Cal. 2000).

Marx v. Globe Newspaper Co. Inc., 15 Mass. L. Rep. 400; 2002 Mass. Super. LEXIS 455; 31 Media L. Rep. 2034 (2002).

McCullagh, D. (2002, Dec. 23). Will This Land Me in Jail? *CNET News.com*. Retrieved April 29, 2004 from http://news.com.com/2010-1028-978636.html

McLachlan, M. (2003, May 9). Free Speech Case A First For UN Body. *Australian Financial Review*, p. 59.

Nelson, S. S. (2001, July). Electronic Rights: Practical Impact of Supreme Court Ruling on Author/Publisher Digital Use Concerns. *Entertainment Law and Finance, 17 (4)* p. 1.

New York Times v. Sullivan 376 US 254, 270 (1964).

New York Times v. Tasini, 533 U.S. 483 (2001).

No Electronic Theft (NET) Act of 1997, Pub. L. No. 105-147, 111 Stat. 2678 (1997) (codified as amended at 17 USC § 101, 18 USC 506(a) and 18 USC 2319).

Northwest Healthcare Alliance v. Healthgrades.com, Inc., 50 Fed. Appx. 339, 341, 2002 U.S. App. LEXIS 21131 (9th Cir. 2002), cert denied 123 S. Ct. 1909 (2003).

Novak T. and Landau, G. (1990, September 9). E. St. Louis Returns Won for Costello. *St. Louis Post Dispatch*, p. 12A.

Pew Center for Internet Research (August 20, 2000). Trust and Privacy Online:Why Americans Want to Rewrite the Rules, Main Report, Section One, 5: Americans Want a Privacy Guarantee. Retrieved May 11, 2004 from http://www.pewinternet.org/reports/reports.asp?Report=19andSection=ReportLevel1andField=Level1IDandID=44

Playboy Enterprises, Inc. v. Frena, 839 F. Supp. 1552 (M.D. Fla. 1993).

Privacy Act of 1974, 5 U.S.C. § 552a; P.L. 93-579, § 3, 88 Stat. 1897, as amended by the Computer Matching and Privacy Protection Act of 1988, P.L. 100-503, § 1, 102 Stat. 2507.

Project for Excellence, The (2004). State of the Media 2004: An Annual Report on American Journalism: Online Intro. 5-9. Retrieved May 7, 2004, from the Journalism.Org Project for Excellence in Journalism Web site: http://www.stateofthenewsmedia.org/index.asp

Pryor, L. and Grabowicz, P. (2001, June 3). Privacy Disclosure on New Sites Low.

Retrieved May 11, 2004 from USC Annenberg Online Journalism Review Web site: http://www.ojr.org/ojr/ethics/1017956628.php

Recording Industry Association of America (2001, April 9). Letter to Professor Edward Felson, Threatening suit under the Digital Millennium Copyright Act. Retrieved May 7, 2004 from http://cryptome.org/sdmi-attack.htm

Rojas, J.P. (2004, April 7). Peers Unhappy with EU Plans on Civil Disputes. *Press Association Limited*, Retrieved Oct. 12, 2004 from Lexis Nexis Academic database.

Schiesel, S. (2003, July 10). Behind a Hacker's Book, a Primer on Copyright Law. *New York Times, Circuits*, p. 6.

Schneider v. Amazon.com, Inc., 31 P.3d 37 (Wash. Ct. App., 2001).

Shetland Times Ltd. v. Wills, 1997 S.L.T. 669 (Sess. Cas. 1996).

Smith v. Maryland, 442 U.S. 735, 743-44 (1979).

Sonny Bono Copyright Term Extension Act of 1998, Pub. L. No. 105-298, 112 Stat. 2827 (1998) (codified as amended at 17 U.S.C. § 108, 203(a)(2), 301(c), 302, 303 and 304).

State v. Heckel, 143 Wn. 2d 824, 832, 24 P.3d 404 (2001) (striking Washington's anti-spam statute pursuant to the Commerce Clause).

Stratton Oakmont v. Prodigy, 1995 WL 323710, 4 (N.Y. Sup. Ct. May 24, 1995), reh'g denied, 1995 WL 805178 (N.Y. Sup. Ct. 1995).

Telecommunications Act of 1996, 47 USC § 222.

Ticket Master Corp. v. Tickets.com, Inc. Case No. CV 99-7654, 2000 U.S. Dist. LEXIS 4553; 54 U.S.P.Q.2D (BNA) 1344, 1346 (C.D. Cal. 2000).

Unintended Consequences: Five Years Under the DMCA. Retrieved Feb. 6, 2004, from the Electronic Frontier Foundation Web site: http://www.eff.org/IP/DMCA/20031003_unintended_cons.php

Universal City Studios v. Reimerdes, 111 F. Supp. 2d. 346 (S.D.N.Y. 2000), *aff'd* Universal Studies v. Corley, 273 F.3d 429 (2d Cir. 2001).

U. S. Constitution, Art. I, § 8, cl. 8.

U. S. Department of Commerce Safe Harbor. Retrieved May 7, 4004 from http://www.export.gov/safeharbor/index.html

United States v. LaMacchia, 871 F. Supp. 535 (D. Mass. 1994).

United States v. Miller, 425 U.S. 435, 443 (1976).

U.S. West, Inc. v. FCC, 182 F. 3d 1224, 1234-35 (10th Cir. 1999).

Viegas, F. (2004, January). Blog Survey: Summary of Findings. Retrieved May 11, 2004 from Massachusetts Institute of Technology Media Lab Web site: http://web.media.mit.edu/~fviegas/survey/blog/results.htm

Weiss, J. (2004, May 10). Blogs Colliding With Traditional Media; Convention Credentials Expected for Web Logs. *Boston Globe*, p. B1.

Yahoo!, Inc. v. La Ligue Contre Le Racism Et L'Antisemitisme, 169 F. Supp. 2d 1181, 1188 (N.D. Cal. 2001).

Young v. New Haven Advocate, 315 F.3d 256, 260 (4th Cir. 2002), cert denied 2003 U.S. LEXIS 3743 (U.S., May 19, 2003).

Zimmerman, D. L. (1999). The More Things Change, the Less They Seem "Transformed": Some Reflections on Fair Use. *Journal of the Copyright Society of USA*, *46 (2)*, 251.

Zippo Manufacturing Co. v. Zippo Dot Com, Inc., 952 F. Supp. 1119, 1124 (W.D. Pa. 1997).

In: Advances in Communications and Media Research, Volume 4 ISBN 1-60021-189-5
Editor: William H. Miller, pp. 19-28

Chapter 2

THE CONVERGENT SOCIETY AND THE MEDIA INDUSTRIES

Barry R. Litman
Michigan State University, USA

ABSTRACT

Weekly newspapers enter the millennium increasing in number and circulation. This is because entry barriers are lower in weekly newspaper markets than in daily newspaper markets. Many weeklies also distribute some or all of their circulation free. Advertisers unwilling to pay prices charged by large dailies may find weeklies attractive. Weeklies may also offer content that strengthens readers' ties to other members of geographic, religious or social communities. The number of weeklies owned by groups is increasing, which may allow those papers to extend scale economies in the production of advertising and news. Newspaper companies that own dailies are acquiring weeklies to extend their reach into new segments of their markets. Weeklies appear well positioned to thrive in the new millennium.

INTRODUCTION

As the information age takes on increasing importance in American society, at the same time and speed we witness the corollary movement of institutions – including those in the media industries – toward convergence. Some would argue that these two trends are inextricably intertwined and mutually reinforcing; that the American market miracle is performing as Adam Smith (19xx) envisioned: the natural interplay of forces work toward equilibrium. Above all else, the process of convergence is a process of achieving a steady state equilibrium, and a stable equilibrium is a highly desirable goal.

The concept of convergence is not new. Along with the concepts of "networking," "digital," and, of course, the "Internet," convergence has framed the transition toward the information society of the new millennium. Convergence has usually meant that the widely divergent technologies would come together and standardize with a common format that would permit a seamless network across various, heretofore, separate platforms. Through digitization, one could easily move from telephony to electronic and home video and to the internet. Voice, image and information bits could freely move and combine in endless directions. Even traditional mass media, such as newspapers and broadcasting, would flow together and share both informational and entertainment content. It was a hope and a dream of a new frontier, a new society which extended the democratic processes to the marketplace of ideas and discourse. The prophets of this new information age and convergent society conveniently tied their dreams to the arrival of the new millennium; hoping to forever emblazon this event in world history.

Convergence as a Path to Total Equilibrium

This last year witnessed the passing of James Hodge, the noted physicist who is credited with the Big Bang theory of how the earth was formed. For Hodge, and others, a giant series of collisions occurred between planets, asteroids and other debris that created the fertile ecosphere that eventually supported the evolution of human life. We can now borrow from this theory in physics and apply it directly to the ongoing convergence of media technologies of present times. This collision of communication media is vividly depicted in Figure 1 which illustrates the collision of three worlds--information[1], communication and entertainment and the accompanying big bang. The intersection of the media industries shows how far convergence has proceeded and what areas of overlap are emerging. A perfect convergence would show one large circle similar to a total eclipse of the sun.

Were this Venn diagram to be equivalent to a total eclipse, then convergence would be complete and the aforementioned global equilibrium would be achieved. However, the diagram only illustrates a partial eclipse which signals that while there are significant areas of commonality, there are also areas of dissimilarity and distinctiveness among the media.

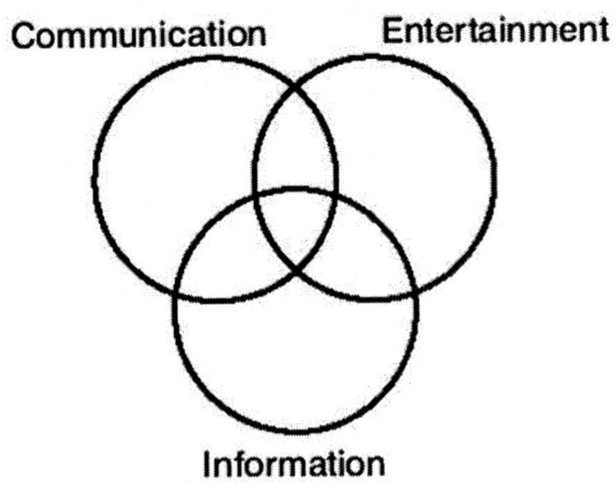

Figure 1. Macro Convergence in Product Space.

[1] Information in this context includes both news and advertising.

To explore this diversity, we then take the macro idea of convergences to the level of each micro sphere. For example, in the entertainment sphere, there is competition for audience attention and advertising revenues among television, VCR's, cable, DBS, newspapers and motion pictures. Each was formerly its own separate industry but, as shown in Figure 2, at the micro level the media industries have begun to blend together as the boundaries become blurred.

For example, the motion picture studios became television networks and VCR distributors and the cable systems began to own stations and networks. This is a clear illustration of Dimmick's biological niche theory (Dimmick, 2003) in which firms now extrapolated to industries fight for resources in the competitive ecosphere, where "survival of the fittest" determines the ultimate winners.

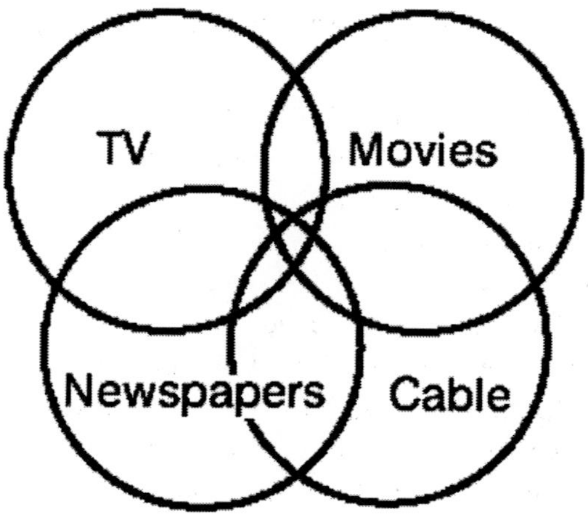

Figure 2. Macro Convergence in Entertainment Space.

The same kind of explanation can be taken to an even lower micro level, namely that of the individual industry. Michael Porter (1985) has described the struggle for market share within an industry as the continual fight to achieve a sustainable market position. This struggle translates into choosing from an assortment of strategies. One choice is price minimization, another maximum product differentiation or alternatively research and development leadership, with the goal to capture and retain market share. Hence, not even a microeconomic equilibrium can be achieved. A partial eclipse continues as firms seek commonality in certain areas and differentiation in others. This leads one to conclude that there are several layers in the struggle for media convergence. Convergence occurs in both macroeconomic and microeconomic dimensions, but, as is discussed below, the improbable goal of achieving a global equilibrium requires a corresponding array of mini-equilibria in all sub-industries.

Convergence through Product Movement

There is another way of conceiving this abstract process of convergence. This explanation suggests that convergence is really the repositioning of firms closer and closer to each other in product space. Michael Porter (1985) discussed this phenomenon of product space with his theory of strategic groups. For Porter, the concept of an overall industry is no longer operational, rather, product space is partitioned into smaller units called "strategic groups" which are really sub-industries themselves.

What becomes crucial in this product formulation is that within each strategic group, there emerges one or two dominant firms that control the industry. These are the General Motors (and Fords) of the industry. In the newspaper industry, Gannett, Knight-Ridder and Thomson are appropriate counterparts. There is a jockeying for position as other firms seek a unique product niche. Once again product differentiation rather than uniformity prevails, and there is a constant struggle between the centrifugal forces that drive firms apart and the centripetal forces that draw them together. The issue of convergence can now be seen in the form of a continuum, with convergence at one end and divergence at the other. The strategic groups and the firms that reside within this continuum must choose the optimal level of convergence and divergence. In this light, it may appear that total convergence cannot be achieved at any level--neither at the micro nor the macro level. Thus, convergence becomes a false god that requires only partial devotion.

THE EFFECT OF TECHNOLOGY

There is yet another destabilizing force that threatens the achievement of full convergence; it is the inexorable march of technology and the havoc it can wreak on industry structures and positions. Joseph Schumpeter (1942) called this whirlwind force a "gale of creative destruction." This gale arrived in the form of new technologies, new forms of business organization and new managerial approaches that obsolesced the status quo and substituted new products for old. The gale was creative and destructive simultaneously and sent a shock wave to those who possessed monopoly power. No firm was immune from these inexorable forces. Unless the dominant companies heavily invested in research and development and were lean and efficient and technologically savvy, their fate would be sealed. Hence, the market could be self-adjusting and restore competitive structures to those that had been corrupted. Such a life threatening force acted as a death knell warning to those firms who sought to lead the "quiet life," free from the anxieties and frenetic pressure of competition.

Technology is volatile and unpredictable. It moves simultaneously in many directions, and, as in the myth of Pandora's Box, once the box has been opened, the forces cannot be contained at all. In an analogy to the tales of the Arabian nights, once the technology genie escapes, it can't be put back in the bottle. In economic terms, the gale of creative destruction causes a period of disquieting disequilibrium that cannot easily be controlled nor easily be understood. The story of the latter part of the last century is replete with examples of new revolutionary technologies arriving on the doorsteps of traditional media and causing exogenous shocks to the equilibrium system. Such revolutions involving television, cable, VCRs, personal computers and the internet all disturbed the status quo and rearranged product space. These technological disturbances have not yet stabilized today's media industries.

Media Consumption and Technological Developments

Maxwell McCombs (1972) studied this march of new technologies longitudinally over a period of fifty years and discovered a remarkable consistency in the share of disposable income devoted to media expenditures. This consistency was similar to the position of necessities of life (i.e., food, clothing, transportation and shelter), leading McCombs to conclude that the media shared the same lofty position. He termed his concept "the principle of relative consistency" (PRC) and suggested that new technologies displaced the old in the form of a zero-sum mathematical game that reinforced the consistency of media expenditures. Further research extended the time frame to the second half of the 20th century and discovered that the aforementioned revolutions were additive not substitutable (McCombs and Nolan, 1992). Rather than a zero-sum game, media consumption and expenditures were a positive-sum game, thus creating even greater importance for this sector of the economy than McCombs had envisioned, especially with the arrival of the new information technologies of the last decade. It is now widely accepted that the information technology (I.T.) sector is the largest and fastest growing across all of American industry.

The lesson to be learned from this portrayal of technology is that even if the technologies portend to be stabilizing by achieving common standards, they unfortunately create disequilibria processes that often are chaotic and unpredictable. Disequilibria are the norm, and equilibria are often short-lived and fragile as new endogenous and exogenous shocks continually appear. For this reason, the global equilibrium situation associated with media convergence may not be easily achievable or maintained in the fast paced technology-progressive information age. Rather, it may be a utopian ideal, a Camelot that appears for only brief instances of time, just like a total eclipse of the sun.

Convergence through Mergers

There is another undeniable and inexorable force that is simultaneously occurring and directly affecting the achievement of media convergence, namely, the trend toward conglomeration of the media. Recent mergers such as AOL-Time Warner and the Comcast – Disney bid illustrate the degree of combination occurring in the media and information sectors. These mergers may take one of two forms: (1) mergers of similar or related products (product extension) or (2) mergers of unrelated products (pure conglomerate). In either form, the mergers illustrate the innate tendency of firms to grow larger and more powerful within and without their central power bases. This "urge to merge" may lead firms to the pursuit of assets that are outside their traditional areas of expertise. While the merger proponents and prophets proclaim many different advantages and positive results from such acquisitions, the real world evidence is hardly supporting or encouraging (Adams and Brock, 1989), and there is a creation of debt that often leads to disaggregation in the final analysis.

In more specific terms, there will be claims of increased efficiencies of all kinds ranging from economies of scale and scope to the elimination of redundant personnel, waste, fraud and abuse. An extra bonus should be the creation of synergy (i.e., 2+2=5) in which the addition of disparate areas creates a whole that supersedes the mere addition of the parts. These super efficiencies may be expected to emerge either from special management skills that can efficiently coordinate many different product lines simultaneously or from the cross fertilization of ideas from one division of an enterprise to another. Hence, a blockbuster movie like *Shrek II*, *Harry Potter* or *Spiderman* causes rippling effects throughout the media value and merchandising chain. Other motives given for such accumulations of assets are the need to achieve a global status to counteract foreign competitors and, most critically, the need to prepare for the new information society by building proper infrastructures. If such mergers better prepare firms to consolidate expertise and resources to take advantage of new opportunities created through convergence, then these benefits could provide additional support and impetus towards achieving the total equilibrium described above.

However, the conglomerate movement has been more commotion than motion! The efficiencies and synergies prophesized seldom materialize, rosy profit predictions sour, and firms are left with heavy debt structures that often cause internal divestiture and dismemberment (recall the plight of ATandT). The largest conglomerate merger of the information age and one of the largest in corporate history occurred several years ago when AOL acquired Timer Warner. This merger has been such a total failure that it has jeopardized the financial standing of AOL, once the darling of the Wall Street community.

The lessons of past conglomerate merger failures have reappeared again as "new wine in old bottles." The attempt to combine disparate enterprises is fraught with problems even to the point of clashing corporate cultures or ego-maniac CEO's. And this poor fit of the conglomerate puzzle is not limited to mergers and acquisitions but applies also to joint ventures where supposedly the financial and experience of two or more companies is brought together to allay and spread risk and take advantage of special expertise from each contributing company. Many such proposed joint ventures are short-circuited before their consummation or else suffer a very short half life. The best examples of this are the Bell-Atlantic-TCI and the Comcast-Disney attempted takeover.

In summary, while the merger and acquisition or joint venture route may be one critical path toward achieving convergence, it is not the miracle salve that cures the wounds. Rather it may be a mere chimera, a sleight of hands that create a false dream and the search for El Dorado.

MEDIA CONVERGENCE AS COEXISTENCE: THE INTERNET

What is then left with the notion of media convergence? Is it a meaningless concept "full of sound and fury" (Shakespeare, 1500), or a whimper rather than a bang? (Eliot, 1936). There appears to be a middle ground that can create a steady state equilibrium around a new epicenter. In the ancient days of the Roman Empire, there was an old saying "that all roads lead to Rome." Literally this wasn't true, but what it signified was that Rome was the epicenter for thought and action. In today's information age, all information and telecommunications lead to the internet – the new epicenter and the super magnet that is the focal point that attracts all media together. It is a pulling force with a compelling prominence that is irresistible. Like all strong magnets, it attracts opposites, which is the true explanation for the coexistence of convergent and divergent forces.

What makes the internet so powerful a magnet is its multidimensional functionalism – its tremendous versatility and utility. It is a great source of news and provides access to a wealth of information. It is an interpersonal communication medium without parallel. It has capability for gaming, gambling and adult entertainment. It can also provide instruction and is becoming an electronic market for commerce, banking, political polling and related activities. These functions underlie the effective demand for internet use by consumers.

The internet also fulfills the critical attributes of every communication technology: speed, distance, quantity, quality and diversity. These five attributes are crucial in determining the adoptability of new services, which is nothing more than assessing the speed at which effective demand moves forward. To the extent that a new technology is superior on providing these characteristic attributes, it is more likely to accelerate its trajectory toward adoption (Rogers, 1962).

Yet as potent a media as the internet is and will be, it is not omnipotent. It has not displaced all the other media and in fact is still a lesser player in the high stakes media game. Insight into this process can be gleaned by examining the demand for media using modern consumer demand theory pioneered by Lancaster (1991). To understand the difference between the traditional and new theories of consumer media demand, we will briefly highlight each approach. Both approaches assume a standard product price.

The Traditional Approach to Media Demand

The traditional approach to media demand suggests that consumers wish to obtain as much utility as possible to satisfy their unquenchable thirst of wants and needs and desires. This utility is embedded in different amounts in a variety of products that are available for consumption. Therefore, it is the product that bears the desirable fruit of utility. The dilemma for consumers is that they demand more products than they can truly afford and, given their limited discretionary income, must make choices that yield the optimal product mix. In other words, consumers must face budget constraints as they desire utility from media. Consumers follow the familiar indifference curve approach (Mansfield, 1975) whereby a map of indifference curves is provided in such a way that the consumer is indifferent along any single curve that provides the same degree of utility but still desirous of moving up to higher levels of utility. Hence we have identical utility along an indifference curve but higher levels of utility as more products become financially available.

Equilibrium is reached where the highest indifference curve is tangent to the budget constraint line. This tangent situation yields the optimal mix of products and then provides a mechanism for changing the price of one product while all others as well as income are held constant. The locus of points that are deduced at different prices trace out a demand curve for a particular product. The same process can then be pursued by the other products.

The New Theory of Media Demand-Media Characteristics

The Lancaster (1991) model reverses polarity and now suggests that products contain a series of "characteristics" that bear the utility that consumers seek. The product is not important itself! It only houses the embedded characteristics. For example, consumers may desire certain information appliances that provide differing amounts of characteristics such as internet access, speed, short and long term memory, portability and other complementary factors. Alternatively, consumers may desire services that provide the characteristics of entertainment, information, communication, education and other related uses. Whether the former are provided by PCs, palm devices, local area networks, etc. is not as relevant as the underling characteristics themselves. Similarly, whether cable, DSL or DBS provides the multichannel dimensionality for the latter characteristics is secondary to the characteristics.

Figure 3 represents this novel approach to understanding media demand. The model highlights three characteristics (or attributes): speed/distance, quantity, and quality and five media: cable, TV broadcasting, newspapers, motion pictures and internet. All five media possess the three characteristics but in different proportions. Points "a" through "e" reflect the different quantities of each medium associated with a standard price. The points are connected to form a trapezoid known as the "characteristics frontier" – the opportunity set of media available to obtain the desired characteristics. The model shows three indifference curves, IS^1 through IS^3.

Overlaying the media characteristics is a map of consumer preferences expressed relative to the characteristics represented by the series of indifference curves (IS^1 through IS^3). Equilibrium is achieved by the tangency of the highest indifference curve with the characteristics frontier. This point reflects the optimal amount of equilibrium characteristics obtained and the mix of media that deliver those characteristics. Of course the shape of the indifference curve (which also represents the marginal rate of substitution) will greatly impact which media are chosen; similarly, the grouping of media characteristics considered can also tip the balance.

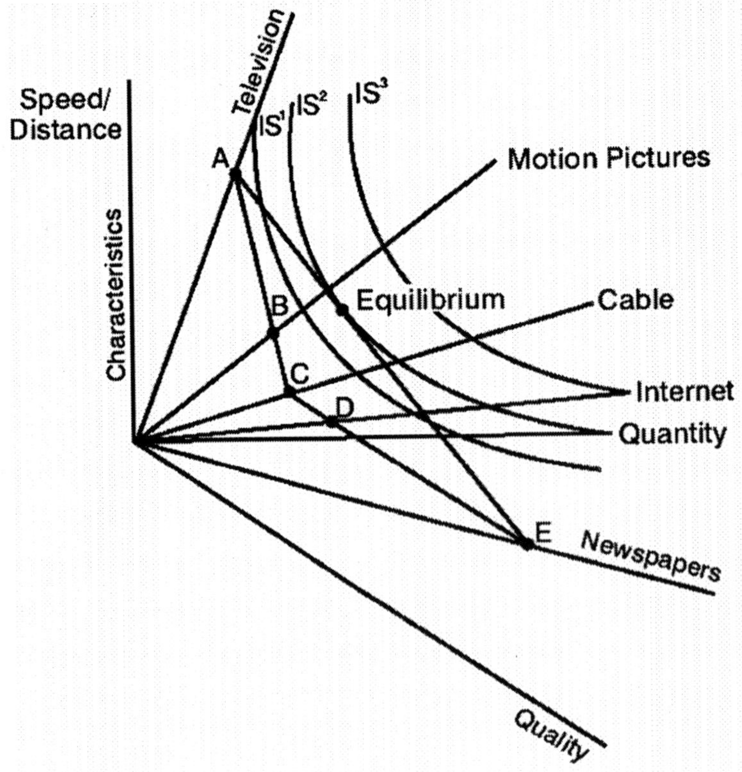

Figure 3. Consumer Equilibrium in Characteristic Space.

In Figure 3, the equilibrium occurs where the second indifference curve is tangent to the trapezoid characteristics frontier. This appears closer to the mass media of motion pictures and cable, indicating that these two media are the right combination to yield maximum utility. Thus, the equilibrium solution need not be a single unique medium, but rather can be a combination of harmonious media. For this instance, one might say that consumers derive utility from consuming movies in theaters as well as from basic and pay cable television.

The New Theory of Media Demand-Media Functions

We can also replace the media characteristics by media functions (e.g., entertainment, information and interpersonal communication) and determine both the consumers' optimal amount of each function and which media are optimal for delivering these functions. If we shift the analysis into multidimensional (N) space, which would consist of X number of media, Y number of characteristics and Z number of functions, we can determine the equilibrium quantity of characteristics, functions and media simultaneously by the intersection of the resulting opportunity ellipse and the indifference surface in the multidimensional, N, space.

Applications of the Model to the Newspaper Industry

The most interesting feature of the characteristics demand model when applied to newspapers is that it is perfectly consistent with traditional academic and industry standards of quality journalism. Today's modern newspapers can still gather and report news the "old fashioned way" and thereby embed quality and utility to their product. Hence, such journalistic concepts as: accuracy, fairness, balance, depth, breadth and speed are still alive and meaningful in the new media millennium. In fact, these concepts of quality newspapers form an array of characteristics as indicated in the Lancaster approach from which consumers derive utility and satisfaction and which form the basis for an effective demand.

Furthermore, it should be recalled that the traditional interrelated circulation and advertising demand models for daily newspapers (Litman, 1988; Litman, 1992) share interaction effects that contribute to multiplier and accelerator effects. Yet, embedded within these equations is a quality variable that also helps to predict consumer demand for circulation and advertiser demand as well. If we now measure newspaper quality by an array of characteristics, these can be converted into scales by factorial analysis. Hence, the Lancaster model can easily fit into the traditional demand relationships.

The financial commitment theory was pioneered by two editors of this volume (Litman and Bridges, 1986) as a proxy for true newspaper quality. When it was thought impossible to achieve a consensus on the subjective ingredients of a quality newspaper, the financial commitment of publishers toward producing the quality newspaper became a second best approach. It was successfully applied to various public policy issues of what the proper market structure should be to insure quality newspapers and what might happen if monopoly conditions appeared (Litman and Bridges, 1986). What this new revival of the Lancaster characteristics model suggests is that the traditional newspaper values outlined above may not be antagonistic to the financial commitment of publishers toward creating a quality newspaper. The two approaches work side by side; the financial commitment facilitates the production of the quality newspaper. They are two sides of the same coin. A return to traditional journalistic values may be the means by which the newspaper industry survives in the competitive jungle.

In today's technologically changing market, the most difficult issue facing the newspaper industry is how to position itself so as to present the newspaper's attributes and functions in the best fashion and at the best possible value to consumers. The industry should emphasize those attributes and functions it can deliver at a comparative advantage without trying to deliver everything. With respect to the internet, newspapers should be both suppliers and demanders of information as they compete with other media for consumer attention. Another alternative is to cooperate with other media to produce a joint product. Most important is the industry's decision of how much newspapers wish to be different. That is, what point in product space they intend to occupy, closer toward convergence or otherwise divergence. This is the most difficult management dilemma of the new millennium for the newspaper industry, and one which has no easy solutions.

References

Adams, W. and Brock, J. (1989). *Dangerous pursuits*. NY: Pantheon Books.

Dimmick, J. (2003). *Media, competition and coexistence: The theory of the niche*. Mahwah, NJ: Erlbaum.

Eliot, T. S. (1936). "The hollow men," in Eliot, T. S. *Collected poems*. NY: Harcourt Brace.

Irwin, M. (1984). *Telecommunications America: Markets without boundaries*. Westport, CT: Quorum Books.

Lancaster, K. (1991). *Modern demand theory*. Aldershot, England: E. Elgar.

Litman, B. (1988). Microeconomic foundations of newspaper research. In McCombs, M.(Ed.), *Concentration and monopoly*, Ablex.

Litman, B. R. (1992). Economic aspects of program quality: The case for diversity. *Studies of broadcasting, 28*,

Litman, B. and Bridges, J. (1986). An economic analysis of newspaper performance. *Newspaper research journal, 7*, 740-44,792.

Mansfield, E. (1975). *Microeconomics*. W.W. Norton.

McCombs, M. (1972). Mass media in the marketplace. *Journalism monographs*, No.24.

McCombs, M. and Nolan, J. (1992). The relative constancy approach to consumer spending for media. *Journal of media economics, 5*, 43-52.

Porter, M. (1985). *Competitive advantage: Creating and sustaining superior performance*. New York: Free Press.

Rogers, E. (1962). *Diffusion of innovations*. New York: Free Press.

Schumpeter, J. (1942). *Capitalism, socialism and democracy*. New York: Harper and Row.

Shakespeare, W. (1500). *Macbeth*, Act V, Scene V.

In: Advances in Communications and Media Research, Volume 4 ISBN 1-60021-189-5
Editor: William H. Miller, pp. 29-51

Chapter 3

THE "*NUESSEL-STEWART AGEISM RATING INSTRUMENT*": LITERATURE REVIEW, FORMAT, IMPLEMENTATION, RECOMMENDATIONS FOR USAGE

***Frank Nuessel*[*] *and Arthur Van Stewart*[†]**
University of Louisville, Louisville, KY, USA

ABSTRACT

Ageism, or discrimination and prejudice on the basis of age, promotes a climate that promotes negative or positive or negative attitudes and beliefs based on chronology. The literature on this subject is vast with much of the focus on negative attitudes and behavior toward older adults. In fact, ageism may be negative or positive and aimed at younger and older people. This paper discusses the literature on ageism as a background for the evaluation instrument known as the "*Nuessel-Stewart Ageism Rating Instrument.*" This assessment vehicle is a form with "yes/no" questions, a numerical scale, Charles Osgood's semantic differential format, and a section for written commentary. The purpose of this instrument is to provide instructors with a format for assessing negative and positive verbal and visual characterizations of ageism across the chronological spectrum. While it is an instrument to assess ageism, it also serves as a pedagogical tool for classroom use and for homework assignments through practical, applied assignments. The purpose of this presentation is four-fold: (1) to discuss the research on ageism; (2) to provide an example of the "*Nuessel-Stewart Ageism Assessment Instrument;*" (3) to exemplify its usage with text and image alone or in combination; and (4) to provide a useful bibliography on this topic.

[*] Department of Classical and Modern Languages; Program in Linguistics; University of Louisville; Louisville, KY 40292-0001; (502) 852-6686 (office); (502) 852-8885 (Fax); fhnues01@athena.louisville.edu
[†] Department of Orthodontic, Pediatric and Geriatric Dentistry, School of Dentistry, University of Louisville, Louisville, KY 40292-0001; (502) 852-1332 (Office); (502) 852-4388 (Fax); avstew01@gwise.louisville.edu

INTRODUCTION

Ageism is an attitude (prejudice) or a behavior, or act (discrimination) which deliberately demeans, disenfranchises, or favors a person or an entire category of people on the basis of their age (see Palmore 1990, 1999 for a full discussion).

Part 1 of this paper reviews the literature on the phenomenon of ageism. Part 2 provides an overview of ageism in the print and non-print media. Part 3 discusses the effects of ageist depictions in the media. Part 4 discusses the verbal manifestations of ageism. Part 5 reviews the non-verbal signs of ageism. Part 6 features an overview of the major factors in an ageism-rating instrument. Part 7 features the "Nuessel-Stewart Ageism Rating Instrument" designed to rate print and non-print materials in terms of their ageist verbal, non-verbal, and iconic content.

AGEISM

Butler (1969) introduced the neologism "ageism." In his now classic article on this topic, Butler (1969: 243) defined this form of bigotry in the following way:

> age-ism reflects a deep-seated uneasiness on the part of the young and middle-aged--a personal revulsion to and distaste for growing old, disease, and disability; fear of powerlessness, 'uselessness,' and death.

Subsequently, in his Pulitzer Prize-winning book, *Why Survive: Growing Old in America,* Butler (1975: 12) would allude to his collaborative volume with I. M. Lewis (Butler and Lewis 1973) on aging and mental health in which he described the phenomenon in the following way:

> Ageism can be seen as a process of systematic stereotyping of and discrimination against old people because they are old, just as racism and sexism accomplish this with skin color and gender. Old people are categorized as senile, rigid in thought and manner, old-fashioned in morality and skills. .... Ageism allows the younger generation to see old people as different from themselves; thus they subtly cease to identify with elders as human beings.

Within a decade of its conception the term ageism became firmly entrenched in our vocabulary as evidenced by its rapid inclusion in the dictionary. A decade after its creation, *The American Heritage Dictionary of the English Language* included this word. That dictionary (Morris 1979: 24), for example, defines this notion as "discrimination based on age; especially discrimination against middle-aged and elderly people."

The practice of ageism, of course, existed long before a term for this behavior was created. Both descriptions of the ageism focus on negative aspects of the treatment of older adults. In their important work on this subject, Levin and Levin (1980) perpetuated this unfavorable delineation of the treatment of older adults, as well as the general portrayal of old age. It was not until the publication of Palmore's (1990) book on this subject that the concept of ageism received a more balanced treatment.

Bytheway (1995: 14) offers the following definition of ageism:

1. Ageism is a set of beliefs originating in the biological variation between people and relating the to ageing process.
2. It is in the actions of corporate bodies, what is said and done by their representatives, and the resulting views that are held by ordinary ageing people, that ageism is made manifest.

In consequence of this, it follows that:

(a) Ageism generates and reinforces a fear and denigration of the ageing process, and stereotyping presumptions regarding competence and the need for protection.
(b) In particular, ageism legitimates the use of chronological age to mark out classes of people who are systematically denied the resources and opportunities that others enjoy, and who suffer the consequences of such denigration, ranging from well-meaning patronage to unambiguous vilification.

Palmore (1990: 4) defines ageism as "*any prejudice or discrimination against or in favor of an age group*" (emphasis in original, FN). The latter definition captures the fact that an entirely negative or positive assessment of a person on the basis of age (young or old) is not objective, hence, inaccurate.

In spite of Palmore's recognition that ageism is not always inimical to the well-being of older adults, much of the research related to the depiction, perception, and treatment of older adults has revealed an overwhelmingly negative depiction of old age, and the older adult. Of course, this may reflect a predisposition on the part of investigators to search for negative factors in the treatment of the old. In fact, Kalish (1979) referred to this possibility in an essay with the partial title of "the new ageism." By this expression, Kalish meant that there is an unwarranted tendency on the part of some gerontological and geriatric professionals to focus on a small minority of older people who are, in fact, dependent, poverty-stricken, and incompetent. A few years, later in the Donald P. Kent Memorial Lecture, Robert Binstock (1983) alluded to Kalish's essay, while trying to debunk some of these same myths surrounding old age, and older adults.

Four Categories of Ageist Misrepresentation

Four basic kinds of ageist misrepresentation of older adults occur (Nuessel 1982). The first is *distortion* which is the assignment of negative physical, behavioral, and mental traits to older adults as a group. This tendency is found verbally in such adjectives as *wrinkled*, *cantankerous*, and *old-fashioned*, *grumpy*. These "words," in turn, are reified in the visual or performing arts as particular kinds of behaviors (Nuessel 1982, 1992c). A second form of media ageism is *degradation* or the systematic characterization of older people as inferior or obnoxious on intellectual or physical grounds. Again, this type of stereotypical depiction occurs through the use of certain descriptive adjectives frequently associated with older people, e.g., *feebleminded*, and *wizen*. Next, *exclusion* constitutes a more subtle form of ageism since older adults do not populate certain environments as if this segment of society does not exist. In effect, this "invisibility" is a kind of segregation that demeans the worth of older human beings in a particularly nefarious way. Finally, *subordination* amounts to the purposeful placement of older adults in demeaning or servile roles, i.e., contexts in which an older person has no meaningful decision-making authority. Finally, the infantilization of an older adult, i.e., treating an older person in the same way that an adult would treat a child, e.g., vocal intonation, a pat on the head, the use of demeaning diminutive terms of putative endearment (*dearie*; see Nuessel and Stewart 1999), is yet another which older adults are demeaned.

The Image of Older Adults in the Print and Non-Print Media

This section will contain a brief overview of the various media-based studies on the image of older adults in the media. This subsection is meant to be representative rather than inclusive (see Nuessel 1992a 1995a for a comprehensive listing of these materials). This introduction to the topic of images of older adults in the media is necessary in order to provide the reader with the prerequisite background necessary for an appreciation of the copious research carried out on the image of the older adult in the media during the past quarter century (Nuessel 1992a,c, 1995a). At the same time, the results of this previous research indicates that ageist depictions of older adults has permeated the print and non-print media. Hence, the ability to decode these sometimes subtle messages is necessary to develop a discerning audience.

In this section, a selective discussion of depictions of older adults in the media will be provided. The print (magazines, newspapers, books), and non-print (cartoons, television, music, art) will be featured. In addition, there are some media in which this traditional dichotomy overlaps, e.g., printed cartoons, and greeting cards, and television because their formats feature both verbal and graphic materials. Nuessel (1992a) contains a comprehensive discussion of all the available materials discussed in this section. Moreover, new technological formats (CD-ROM, music video) likewise feature an admixture of print and iconic materials. The Internet accessed by a variety of proprietary programs is yet another medium, which contains a combination of print and visual elements.

Children's Literature

Numerous studies have been carried out on the role and the depiction of older adults in children's first literature (e. g., Ansello 1976a, b, c, 1977, Blue 1978, Meadows 1986, Seefeldt, Galper, Serock, and Jantz 1978). Results of these research projects indicate that older people are often mis-represented. Because these books represent a child's first exposure to a socially constructed reality (Berger and Luckmann 1966), their contents may have a long-term attitudinal effect on young readers. Another domain for research is that of the "fairy tale" with its classic "boogey man" who is frequently old with all of the stereotypic physical and behavioral signs associated with older adults.

Adolescent Literature

A few studies also exist which consider literature designed for the adolescent (e. g., Gerlach 1988, Peterson and Karnes 1976). Again, the older person has a marginalized role and stereotypic image. Continuous exposure to this characterization of the older adult may produce a lifelong distorted image of aging and older people.

Literature

The use of the term literature for this subdivision would seem to suggest that children's literature, and adolescent literature are somehow distinct from other forms of literature. There is clearly a qualitative judgment involved in such evaluations, which remains beyond the scope of this paper.

During the past decade, university-level courses devoted to the theme of aging in literature have emerged across the country in increasing numbers in recognition of the significance of this topic. In fact, at the annual 1988 meeting of the Association for Gerontology in Higher Education, an entire session was devoted to a discussion of these courses as a significant new development in approaches to gerontology (Nuessel 1988, 1995b; Nuessel, Stewart and Cedeño 2001; Nuessel, Makris and Stewart 2002). Several bibliographic reference works feature sections devoted to a literature of aging. Yahnke and Eastman (1990, 1995), for example, have even produced substantial annotated volumes on the literature about aging. In addition to the various journal articles devoted to diverse aspects of aging in literature (Nuessel 1992a, chapter 8), several important anthologies of literary criticism now exist (e.g., Bagnell and Soper 1989, Falkner and DeLuce 1989, Porter and Porter 1984, Woodward and Schwartz 1986) to name but a few.

Art

Artistic renderings of aging and older people provide the researcher with a historical iconography of perceptions of old age in different epochs. This form of expression offers a visual account of earlier viewpoints on the aging process, and the status of older people. At least three important volumes with accompanying graphic illustrations on this topic now exist (Achenbaum and Kusnerz 1978, Covey 1991, McKee and Kauppinnen 1987). These books provide some significant insights into the history of aging from a unique iconic perspective.

Humor

An analysis of the sources of humor in a particular society often provides important insights into that culture's phobias since wit, in its various formats, represents a healthy psychological reaction to collective fears such as the fear of aging. Palmore (1971) first investigated this source to ascertain the basic subject matter of the humor of old age. Among the common themes of such humor are longevity, physical appearance, obsolescence, sexuality, age concealment, old maid, retirement, mental acuity, state of felicity, and death. The significance of this topic was highlighted by the publication of an entire anthology dedicated to the subject (Nahemow, McCluskey-Fawcett, and McGhee 1986). Moreover, studies on related matters such as cartoons (Smith 1979), and greeting cards (Demos and Jache 1981, Dillon and Jones 1981, Nuessel and Stewart 1989) have also appeared.

Newspapers

A limited amount of research into the depiction of older adults in newspapers exists. Extant studies suggest that coverage of older people is scarce (Buchholtz and Bynum 1982), or that little change in quantity and quality of coverage had occurred (Wass, Hawkins, Kelly, Magners, and McMorrow 1985). These findings suggest that additional study of journalistic coverage of older adults is necessary, especially during the last decade.

Magazines

The critical scrutiny of the depiction of aging and older people in magazine formats, thus far, has been selective, i.e., research has dealt selectively with certain modern magazines (Kent and Shaw 1980, Wang 1988), or in magazines from an earlier historical period (Murphy 1980, Range and Vinovskis 1981). Analyses of publications from an earlier era provide a useful timeline to assess changes in the perceptions of aging.

Advertising

Commercial advertisements often project a reflection of commonly held social images of older people. Therefore, these ubiquitous cultural artifacts measure to a certain extent a collective viewpoint about aging and older adults in contemporary society. In the few studies that exist on this topic, it is apparent that older people are either invisible (Gantz, Gartenber, and Rainbow 1980), or they are characterized pejoratively (Smith 1976). Nevertheless, in certain circumstances (advertising aimed at older people), this is not always the case (Kvasnicka, Beymer, and Perloff 1982). There is anecdotal evidence of more advertising targeted at older adults since there is increasing awareness of this age cohort as a group that will generate significant revenue.

Music

The role of popular music as a database to determine attitudes toward aging is significant for the contemporary period (Kelley 1983, Leitner 1983), or other historical eras (Cohen and Kruschwitz 1990). Because this form of popular culture is rapidly dispersed orally in a given culture, the sentiments toward aging and older people will receive widespread diffusion.

Film/Videotape/DVD

Popular films in which older characters are portrayed with compassion, understanding for the older person's concerns and problems are appearing more frequently, e.g., "Harold and Maude," "Harry and Tonto," "Cocoon," and "Driving Miss Daisy," to name but a few examples of this genre. This category film needs to be distinguished from the films that deal with older people from a clinical or gerontological perspective., e.g., the type reviewed in the audiovisual review section of *The Gerontologist.* A relative paucity of research exists in this domain with only a few such studies currently available (Butsch and Baron 1980, Stoddard 1980, Yahnke, 1996a,b).

Television

The impact of television on U.S. society in the past forty years has been formidable. This form of mass communication permeates our daily existence, as has no other medium. Consequently, its role in shaping public opinion, influencing attitudes, and creating a view of the world is inestimable. It is for these reasons that continuous monitoring of, and research into the interface between television broadcasting and attitudes toward aging and older people must be sustained. A number of scholarly inquiries into television's portrayal of older adults and aging are already available (e.g., Davis and Davis 1985 and the studies cited therein, Kubey 1980).

Effects of Ageist Depictions in the Media

The empiricist approach to epistemology holds that we acquire information through sensorial contact with our environment (Danesi 1986, 1989a, b, 1993, 1995, Nuessel 1991, 1992b, 1993, 1994a,b, 1995c). Danesi (1987: 380-381, 1988: 19) has provided a useful sketch of the differentiated functions of the human brain based on Edwards (1989) bimodal typology which specifies in which hemisphere of the human brain certain basic linguistic stimuli are stored and processed. The left mode, or left cerebral hemisphere, and the right mode, or right cerebral hemisphere of the brain are terms employed by Edwards (1989) to refer to the specific operations carried out by each half of the brain. Visualization, the comprehension of prosodic features (pitch, intonation, stress), and synthetic thinking are R-mode features, while verbal memory, analytical thinking, decipherment of meaning are a few of the L-mode functions of the brain. Although Danesi's research has been applied to second-language acquisition, these same epistemological functions also refer to the way in which we acquire prejudice from our cultural milieu.

Because we are bombarded daily from our earliest years with an immense amount of media-generated information that stimulates our left-, and right-cerebral hemispheres, some negative ageist data will infiltrate our brain sensorially and this penetration may result in the creation of negative attitudes toward older adults and aging.

Ageism, like other forms of prejudice and discrimination, is acquired, rather than innate. The source of this bigotry is to be found in our social and cultural environment, and in our institutions. Levin and Levin (1980: 85-86, see also Palmore 1990, 1999) note in this regard that:

> As a cultural phenomenon. ageism is learned early in the life of a child beginning with experience with his parents and later his peers, teachers, and the mass media, and maintained throughout the life span.

Prejudice and bigotry are learned behaviors, acquired from our culture, institutions, and intimate personal associates. Because intolerance and provincialism are not innate, these forms of conduct can be eliminated through appropriate socialization strategies. At this juncture, a few recommendations are in order.

Palmore (1990: 167-175; 1999) provides an enumeration of some worthwhile strategies for change. Activities related to media-based ageism are worth noting here. Individuals, for example, can engage in the following activities: (1) boycott products of companies that engage in ageist practices, or produce ageist advertising; (2) join organizations that fight ageism, and (3) communicate in writing with companies and organizations about ageism and age-related discriminatory practices to name but a few such positive activities. Organizations may also participate in anti-ageist pursuits: (1) disseminate factual information on aging systematically; (2) publish materials that discuss ageism and appropriate strategies for combating this form of bigotry; (3) boycott products of companies that engage in ageist activities.

Anti-Ageism Guidelines

One of the most significant media-related areas of public education involves the widespread dissemination of guidelines for the correct and appropriate depiction of older adults in print and non-print formats. Such an approach is indirect since this strategy is aimed at the authors of media-based materials ultimately intended for mass consumption. Since several such guidelines already exist (Gamse 1984, Mathiason 1977, Wisnieski, Star, and Hermann 1979), this task is facilitated. Moreover, such a preventative strategy strikes at what many gerontologists have identified as the heart of the problem, namely, early acculturation into an ageist mindset through media materials aimed at children in their most formative stages of development.

The anti-ageism guidelines just noted constitute an initial, primitive form of a rating instrument because these pamphlets and lists are informational and educational in their intent. The "*Nuessel-Stewart Ageism Rating Instrument*" (see below), however, is intended to provide a step-by-step procedure for assessing and identifying the various ageist elements in a particular document.

Verbal Communication

Linguistic behavior, or language, is perhaps the most important way of engaging in ageism. Verbal communication may involve oral or scripted communication about older adults, or it may involve conversational dyads or direct communication with older adults. For an overview of verbal ageism see section 6 below.

Ageist Communication about Older Adults

Ageist linguistic conduct is exhibited in communication about older adults (see Nuessel 1982, 1984, 2002). This sort of behavior manifests itself through the use of any of the copious nouns that refer to older adults, either male (*geezer*, *gaffer*, and so on) or females (*witch*, *harridan*, and so forth). In addition, the ageist verbal depiction of older adults in spoken or written communication may employ any of a number of unflattering adjectives (*senile*, *feebleminded*, and so forth), adverbial expressions (*in a senile manner*, *foolishly*, and the like), or through the use of certain predicates that purport to describe popular notions of older adult activity (*drool*, *teeter*, etc).

Ageist Communication with Older Adults

Certain ageist linguistic practices may also occur through direct communication with an older adult. Dialogue between younger and older adults may demonstrate overt ageist language through the use of pejorative vocabulary items such as those enumerated in Nuessel (1982, 1984). More subtle ageist communication may occur in grammatical usage such as artificially simple sentences not unlike the phraseology employed by parents to engage infants in their first language-learning experiences. Such usage, of course, constitutes what Arluke and Levin 1989, see also Gresham 1976) call the infantilization of old age. Other syntactic manifestations of a subtle linguistic ageism include the use of direct commands and tag questions, again, which are strategies commonly employed to reprimand children. The use of first names, infantilizing expressions, and third-person references (in the presence of the party being discussed) are additional forms of demeaning linguistic conduct. Finally, the control of the topics of conversation is an indirect form of domination and subordination. Examples of this type of verbal activity include superficiality of subject matter (weather), intimate topics, and excessive reference to the past.

Non-Verbal Communication

Non-verbal communication includes the following three general domains:

1. Kinesics.
2. Paralanguage.
3. Proxemics.

Subtle, powerful and significant messages can be projected through these channels that coincide with verbal interaction. The non-verbal characterization of older adults may be encountered in a variety of formats. In their purest form, these images consist of visual depictions of older adults, e.g., paintings, sculpture, photographs, and other iconic manifestations (Achenbaum and Kusnerz 1978, Covey 1991, McKee and Kauppinen 1987). Often, however, so-called non-verbal portrayals of older people in a visual format are, in fact, a mixture of both verbal and iconic elements. The verbal component in television, for example, derives from scripted materials and this element co-exists with the orally communicated speech. There may, in fact, be written verbiage superimposed on the television screen simultaneous with the oral message. For an overview of non-verbal ageism, see section 6 below.

Kinesics

Kinesics refers to significant, or meaningful body movement. Key (1975: 10) defines kinesics as the "articulation of the body, or movement resulting from muscular or skeletal shift." Furthermore, Knapp (1978: 12) observes that "body motion, or kinesic behavior, typically includes gestures, movements of the body, limbs, hands, heads, feet and legs, facial expressions (smiles), eye behavior (blinking, direction and length of gaze) and posture." Kinesics may convey substantial collateral information about a speaker's mental, emotional, or physical health. A person's gait, posture, and facial expressions are all forms of non-verbal communication that provide clues about an individual's physiological and psychological status.

Ekman and Friesen (1969) have developed a categorization of kinesics with five separate categories:

1. Emblems.
2. Illustrators.
3. Affect displays.
4. Regulators.
5. Adaptors.

In brief, *emblems* are "nonverbal acts which have a direct verbal translation or dictionary definition, usually consisting of a word or two or a phrase" (Knapp 1978: 13). The "victory" and "peace" signs exemplify this form of kinesic behavior. *Illustrators* are "nonverbal acts which are directly tied to, or accompany speech-- serving to illustrate what is being said verbally" (Knapp 1978: 15). Nodding one's head in affirmation is an example of this nonverbal conduct. *Affect displays* are "primarily facial configurations which display affective states (Knapp 1978: 16). This form of automatic behavior signifies states such as fear, fatigue, happiness, and anger. *Regulators* are "nonverbal acts which maintain and regulate the back and forth nature of speaking and listening between two or more interactants" (Knapp 1978: 16). Diminished eye contact, meant to indicate the termination of conversations, is an example. *Adaptors* may be subdivided into three categories of behavior. The first is the *self-adaptor*, i.e., manipulations of one's own body--for example, holding, rubbing, squeezing, scratching, pinching, or picking oneself" (Knapp 1978: 17-18). The second is *alter-adaptor*, i.e., the attacking and provoking. The last is the object adaptor which "involve[s] the manipulation of objects and may be derived from the performance of some instrumental task--for example, smoking, writing with a pencil, and so on" (Knapp 1978: 17-18).

Paralanguage

Knapp (1978: 18) states that paralanguage is "how something is said and not what is said." Paralanguage encompasses three subcategories. The first is the vocal characterizer (laughing, crying, sighing, and so forth). The second is the vocal qualifier (intensity, pitch, length, and extent). The last is the vocal segegrate (pause indicators such as "um," and so forth).

Proxemics

Proxemics is "the study of our use of perception and personal space (Knapp 1978: 19). One example of this is cultural in nature. For most Anglophones in the United States, the usual, comfortable distance between the speaker and the hearer is approximately fourteen inches. For Hispanophones, this distance is considerably closer, often about six inches. An Anglophone may violate this spatial barrier or "comfort zone" if he or she is known to suffer a hearing or visual deficit. Culture and intimacy determine the prevailing proxemic norms.

In the auditory realm, paralinguistic ageism occurs when older people are customarily characterized by high-pitched, halting speech, punctuated frequently by coughing spells, or uncontrollable wheezing, or other audible vocal characterizations. Research (Benjamins 1986, Ptacek and Sander 1966, Ramig 1986, Ryan and Capadano 1978, Shipp and Hollien 1969), nevertheless, has shown that the voice of an older adult is distinctive and identifiable, yet this identification is not based on stereotypic vocal features.

The significance of non-verbal communication is that this kind of linguistic behavior is learnable as evidenced by creative artists who incorporate such vocal or kinesic modifications into their dramatic performances, especially when a younger actor impersonates an older character. In scripted material, non-verbal behavior is frequently included in the stage directions of plays, or in parenthetical notes about a given protagonist. These special qualities may subsequently be "translated" into a theatric rendering of the fictitious character.

Iconic Representation

The amount of personal information transmitted non-verbally is substantial. Research indicates that only a small amount of a message is conveyed verbally (Kirch 1979). This means that the hearer *qua* observer of the media derives a great deal of information from secondary, non-verbal sources (see Nuessel 1992c for a detailed discussion).

In the visual domain, for example, certain culturally defined, non-verbal stereotypes about old age and older adults exist. These erroneous misconceptions may involve the depiction of older people as habitually dependent upon canes, or ambulators, hearing aids, and other medical paraphernalia for support. Another kind of ageist image is the traditional depiction of older adults as "stooped by age," or adorned in clothing that is "old-fashioned," or "out-of-date." By implication, this garb signifies that the persons themselves are also obsolete, and *passé*. Proxemic stereotyping occurs when an older person is seen violating cultural rules of spatial proximity to hear someone more clearly, or to see a person or an object better. It is often the case that such proxemic behavior is accompanied by kinesic gestures such as "cupping the ear," or "squinting the eyes" to signify the particular physical afflictions of a hearing or visual deficit.

FACTORS IN AN AGEISM RATING INSTRUMENT

Proper use of the "*Nuessel-Stewart Ageism Rating Instrument*" requires the categorization, enumeration, and exemplification of the verbal, non-verbal, and iconic factors involved in ageism (see Ryan, Hummert, and Boich 1995: 154, Nuessel 1992c). This section of the "*Nuessel-Stewart Ageism Rating Instrument*" constitutes an educational preamble to be used as a reference for the evaluator. This section should be presented in conjunction with a discussion of this topic of ageism in a class, a single public lecture, or a workshop on this subject. Moreover, this component should involve the actual evaluation of selected examples of the media depiction of older adults with appropriate discussion. In other words, the usage of this instrument requires the user to have background about its proper usage. This background information and the "*Nuessel-Stewart Ageism Rating Instrument"* will permit raters to deconstruct various print and non-print documents in terms of their possible ageist content.

I. Verbal Communication

A. The Lexicon.

1. Monosyllabic lexical items.
2. Terms used for children ("Dear," Baby," etc., see Arluke and Levin 1984, Gresham 1976; Nuessel and Stewart 1999).
3. Use of subordinating diminutives expressions ("dearie").
4. Use of inclusive expressions ("We are going to take a bath now").

B. The Grammar.

1. Simple sentences rather than complex sentences.
2. Imperatives ("Come along, now").
3. Tag questions ("It's time to go to bed now, isn't it?").

C. Forms of Address.

1. Use of first name especially with a diminutive morpheme (Annie, Tommy, etc.).
2. Infantilizing terminology ("honey").
3. Third person reference in the presence of the older adult being discussed ("He's having a problem today").

D. Topic Management.

1. Limitation of topics for discussion (inappropriately intimate topics, excessive reference to the past, superficial topics, commands to do specific behavior).
2. Frequent interruption when older adult is speaking.
3. Excessive praise for performing common tasks.

II. Non-Verbal Communication

A. Kinesics.

1. Emblems.

The use of a sign with a specific meaning to indicate approval or disapproval of a person's behavior.

2. Illustrators.

a. The nod of the head to indicate permissive approval of an older person's course of action.

b. The back-and-forth motion of the head to indicate non-compliance with a command or suggestion.

3. Affect Displays.

a. Minimal eye contact with an older adult (= disinterest in the person and what he or she is saying or doing).

b. Staring.

c. Rolling one's eyes to indicate disapproval of an older person's behavior.

4. Regulators.

a. Diminished eye contact to indicate subtly a desire to terminate the conversation.

b. Looking at another person or other parts of the conversational space to indicate disinterest in the topic of the other person.

5. Adaptors.

a. Self-adaptor such as twiddling one's thumbs to indicate discomfort with talking to an older person.

b. Alter-adaptor such as patting a person on the head as a sign of approval.

c. Object-adaptor such as fiddling with an object to indicate boredom or disinterest.

B. Paralanguage.

1. Vocal Characterizer.

a. The use of a sigh to indicate displeasure.

b. The use of laughing tone to indicate derision of an older person's demeanor.

2. Vocal Qualifier.

a. High-pitched tone.

b. Unnatural intonation.

c. Loudness.

d. Excessive elongation of certain words.

3. Vocal Segregate.

The use of a pause indicator as a ruse to pretend that you are unsure of what you want the older person to do.

C. Proxemics.

1. Standing too close to a person (invading personal space).
2. Standing over a person in a wheel chair or in a bed (domination and subordination).
3. Standing at a distance as a sign of disinterest or aloofness.

III. Iconicity

An image (painting, drawing, photograph, film, televised program, video cassette, DVD, video game, CD-ROM, Internet communication, etc.) which depicts and conveys non-verbal communication (see II above) or combines images with scripted words (the monologue or dialogue of an actor) or the combination of a written verbiage with an icon.

THE "NUESSEL-STEWART AGEISM RATING INSTRUMENT"

The following "*Nuessel-Stewart Ageism Rating Instrument*" is a procedural model designed to provide evaluators of print and non-print materials by means of a mechanism for determining the ageist content of a specific document and with a device to evaluate the amount of that ageist content. This instrument contains four parts. Part I provides a form for categorizing and documenting a particular characteristic. Part II contains a series of yes/no questions intended to ascertain if the document is print only, non-print only or a combination of both elements. Part III contains a semantic scale of antonymous positive and negative adjectives to assess the content of the document. Part IV features a numerical scale to determine the frequency of instances of ageism represented in a specific document. Part V provides a section for a narrative overview of the document based on the separate, componential analyses of parts II-IV. The purpose of this document is to provide educators with an instrument to help to educate students in the decoding of various print and non-print documents that they encounter in everyday life.

Semantic Meaning

Section 3 below deals with semantic meaning of print and/or non-print documents to be reviewed. Nearly four decades ago, Charles Osgood and his associates (Osgood, Suci, and Tannenbaum 1957, see also Snider and Osgood 1969) produced an influential psycholinguistic volume entitled *The Measurement of Meaning*. That technique involved the use of descriptive adjectives with antonymous meanings. Although we will not employ the statistical apparatus of the Osgood Semantic Differential Measurement Instrument (Osgood, Suci, and Tannenbaum 1957: 76-124). Rather, we are using adjectival antonyms as a way of determining general attitudes and reactions to verbal, non-verbal, and visual stimuli about older adults. We have included a set of exemplary antonymous pairs of adjectives in which the first is a positive trait and the second a negative characteristic. People who use our assessment instrument will be asked to check one of the two adjectives in a column. Fig. 1 exemplifies the types of adjectives used (Chapman 1977, Lewis 1964, Osgood, Suci, and Tannenbaum 1957, Snider and Osgood 1969).

If the stimulus is a written passage, the evaluator will indicate the types and numerical quantity of descriptive adjectives used, e.g., 'happy,' 'sad.' Other negative or positive linguistic features may include predicates that allude to personal conduct, e.g., 'drool,' 'teeter,' and so forth, or, adverbs or adverbial phrases that modify predicates such as 'feebly,' 'in a rambling manner,' and so forth.

Positive Adjective:	check if applicable:	Negative Adjective:	check if applicable:
happy		sad	
Total Number of check marks for column 2 and column 4			

Figure 1. Semantic Differential Measurement.

At the end of the document (section 5), the rater will be asked to write an analytical narrative account about the stimulus (passage or icon) to describe the ageist or non-ageist features of the documented examined.

The "Nuessel-Stewart Ageism Rating Instrument"

"Nuessel-Stewart Ageism Rating Instrument"

I. Nature of the Document to be Evaluated.

A. Document Type:

1. Print Document
   - __ a. Book
   - __ b. Magazine
   - __ c. Journal
   - __ d. Newspaper
   - __ e. Handout
   - __ f. Brochure
   - ___g. Other (Specify)___________
2. Non-Print Document
   - __ a. Film
   - __ b. CD-ROM
   - __ c. Musical recording (vinyl, compact disc, video, DVD)
   - __ d. Video game
   - __ e. Television program
   - ___f. Other (Specify)___________

B. Reference Documentation:

1. *Print Document:*
   - a. Title of Print Document (book, magazine, journal):_______________________________
   - b. Date of Publication of Print Document:_____________
   - c. Page Numbers of Print Document:_________________
   - d. Publication Information of Document (Place, Publisher):________________________
2. *Non-Print Document:*
   - a. Title of Non-Print Document (film, recording, compact disc, video game, DVD):_____________________________
   - b. Date of Non-Print Document:__________________

c. Production Information for Non-Print Document (place of production, producer):______________________________

II. Yes/No Questions

Print (text in a newspaper, magazine, book, scripted materials).

Yes ____ No _____

B. Non-print

Auditory (text, recording of scripted materials).

Yes ____ No _____

Iconic (illustration, photograph, cartoon).

Yes ____ No _____

C. Combination of print and non-print.

Video (icon + sound + text), e.g., television (icon + sound + text), video-cassette, DVD, cartoon

Yes ____ No _____

Audio (sound only based on scripted material) compact disc recording, radio program

Yes ____ No _____

III. Semantic Scale

In the scale in Fig.2, items 1-6 deal with overt physical traits while items 7-15 address behavioral characteristics.

Positive Adjective:	√ if applicable:	Negative Adjective:	√ if applicable:
1. strong		1. weak	
2. attractive		2. ugly	
3. healthy		3. sickly	
4. active		4. passive	
5. well-groomed		5. unkempt	
6. fast		6. slow	
7. sociable		7. unsociable	
8. graceful		8. awkward	
9. pleasant		9. unpleasant	
10. kind		10. cruel	
11. exciting		11. dull	
12. happy		12. sad	
13. dominant		13. submissive	
14. interesting		14. boring	
15. agreeable		15. grumpy	
Total Number of check marks (√) for column 2 and column 4			

Figure 2. Semantic Differential Measurement of Ageism.

IV. Numerical Scale

The following numerical scale will be used in this section. The scale is based on an actual count of the instances of ageist references in a document. This part of the evaluation should be based on the following factors: text and icon. The rater can count actual instances of lexically ageist terms. In those documents with images only, the evaluator may count the iconic stereotypes such as kinesic ("cupping the ears" as a gesture of a hearing deficit), proxemic (holding a written document close to the eyes to indicate vision deficit), and mechanical ageist references (use of thick glasses to signal a vision deficit, or the use of a cane to indicate ambulatory problems). The rater should, therefore, consider the following factors in making this evaluation:

1. Verbal ageism (unflattering lexical references: nouns, adjectives, verbs, see Nuessel 1982, 1984, 2002).
2. Non-Verbal ageism (kinesic, paralinguistic, proxemic references).
3. Iconic ageism (graphic depictions of older adults in an ageist context).

In the actual evaluation, the evaluator needs to maintain a count of the items above by use of a highlighter pen or some other marking device. The rater should indicate the actual number of instances of identifiable discrete ageist items. In this way, if the rater indicates a "5" on the scale below, that person will have maintained an accurate count of the instances of ageism in the document under evaluation.

1 2 3 4 5

---

non-ageist > somewhat ageist > ageist > more ageist > most ageist

1 = No documented ageist content (no verbal, non-verbal, or iconic imagery).
2 = Single documented instance of ageist verbal, non-verbal or iconic imagery.
3 = Two documented instances of ageist verbal, non-verbal or iconic imagery.
4 = Three documented instances of ageist verbal, non-verbal or iconic imagery.
5 = More than three documented instances of ageist verbal, non-verbal or iconic imagery.

This document may be rated on the following categories (N/A = not applicable).

___ Verbal content (actual number of instances___)
___ Non-verbal content (actual number of instances___)
___ Iconic content (actual number of instances___)

Written Commentary.

Describe your reaction to the document (text, icon, combination of textual and iconic material). In preparing this written commentary, the rater should describe the type of text (see section II above). Next, the rater should comment on the qualitative aspects of these instances. Moreover, the rater should refer to the numerical rating of the text (see section IV above). The rater should specify number of ageist instances in the document. Finally, the rater ought to comment on the semantic impact of the document (see section III above).

This narrative should also call attention to the accuracy of the portrayal of the older adults in the document under review by referring to empirical research such as demographic studies that address statistical information on longevity of the older population by race, gender, and so forth. Likewise, the rater should comment on the accuracy of the document with reference to illness, poverty, and other factors. The accuracy issues (Palmore 1990, 1999) should address both excessively positive and negative depictions. The rater should base these judgments on available empirical research.

____

____

____

____

____

Concluding Remarks

This study has provided a discussion and definition of the concept of ageism. The second section reviewed previous research on the depiction of older adults in the print and non-print media. The third section discussed the potential effects of ageist depictions of older adults. This same section discussed previous attempts at media guidelines designed to address ageism. These guidelines constituted a sort of primitive ageist rating instrument. The fourth section discussed and exemplified the verbal manifestations of ageist linguistic behavior. The fifth section defined and demonstrated non-verbal ageist conduct. This section also addressed the iconic representations of non-verbal ageism. The sixth section provided an overview of the various verbal, non-verbal, and iconic properties of ageism with selected examples. In this section, the authors advocated that the "*Nuessel-Stewart Ageism Rating Instrument*" be used in conjunction with an educational introduction to the topic of ageism, and the distribution of the exemplary factors of an ageism rating instrument. This recommendation is intended to ensure that raters base their evaluations on an educated review of materials they evaluate. The final section contains the actual "*Nuessel-Stewart Ageism Rating Instrument.*"

References

Achenbaum, W. A., and P. A. Kusnerz. 1978. Images of aging in America: 1790 to the present. Ann Arbor, MI: Institute of Gerontology, The University of Michigan/Wayne State University.

Ansello, Edward F. 1976a. Ageism in picture books, part I: how older people are stereotyped. Interracial Books for Children Bulletin 7: 4-6.

____1976b. Ageism in picture books, part II: the rocking chair syndrome. Interracial Books for Children Bulletin 7: 7-10.

____1976c. Ageism in picture books, part III: old age as a concept. Interracial Books for Children Bulletin 7: 6-8.

____1977. Age and ageism in children's first literature. Educational Gerontology 2:255-274.

Arluke, Arnie, and Jack Levin. 1984. Another stereotype: old age as a second childhood. Aging # 346: 7-11.

Bagnell, v. D. P., and P. S. Soper. 1989. Perceptions of aging in literature: a cross-cultural study. Westport, CT: Greenwood Publishing.

Benjamins, B. J. 1986. Dimensions of the older female voice. Language and Communication 6: 35-45.

Berger, Peter L., and Thomas Luckmann. 1966. The social construction of reality: a treatise in the sociology of knowledge. New York: Anchor Doubleday Books.

Binstock, R. H. 1983. The aged as scapegoat. The Gerontologist 23:136-143.

Blue, G. F. 1978. The aging as portrayed in realistic fiction for children 1945-1975. The Gerontologist 18: 187-192.

Buchholtz, M., and J. Bynum. 1982. Newspaper presentation of America's aged: a content analysis of image and role. The Gerontologist 22: 83-88.

Butler, Robert N. 1969. Age-ism: another form of bigotry. The Gerontologist 9:243-246.

____1975. Why survive? Being old in America. New York: Harper Torchbooks, Harper and Row Publishers.

Butsch, R. and A. Baron. 1980. Reviewing the reviews: a note--the portrayal of elderly in films and reviews. The Gerontologist 20: 602-603.

Bytheway, Bill. 1995. Ageism. Buckingham, UK/Bristol, PA: Open University Press.

Chapman, Robert L. 1977. Roget's international thesaurus. 4th ed. New York: Harper & Row.

Cohen, E. S., and A. L. Kruschwitz. 1990. Old age in America represented in nineteenth and twentieth century popular sheet music. The Gerontologist 30: 345-354.

Covey, H. 1991. Images of older people in Western art and society. New York: Praeger Publishers.

Danesi, Marcel. 1986. Language and the origins of human imagination. New Vico Studies 4: 45-56.

____1987. Practical applications of current brain research to the teaching of Italian. Italica 64: 377-392.

____1988. Neurological bimodality and theories of language teaching. Studies in Second Language Teaching 10: 13-31.

____1989a. The neurological coordinates of metaphor. Communication and Cognition 22: 73-86.

____1989b. The role of metaphor in cognition. Semiotica 77: 521-531.

____1990. Thinking is seeing: visual metaphors and the nature of abstract thought. Semiotica 80: 221-237.

____1993. Vico, metaphor, and the origin of language. Bloomington, IN: Indiana University Press.

____1995. Giambattista Vico and the cognitive science enterprise. New York: Peter Lang.

Davis, R. H., and J. A. Davis. 1985. TV's image of the elderly: a practical guide for change. Lexington, MA: Lexington Books.

Demos, V. and A. Jache. 1981. When you care enough: an analysis of attitudes toward aging in humorous birthday cards. The Gerontologist 21:209-215.

Dillon, K. M., and B. S. Jones. 1981. Attitudes toward aging portrayed in birthday cards. International Journal of Aging and Human Development 13(1):79-84.

Edwards, B. 1989. Drawing on the right side of the brain. Rev. ed. Los Angeles, CA: Tarcher.

Ekman, Paul, and Wallace V. Friesen. 1969. The repertoire of non-verbal behavior: categories, origins, usage, and coding. Semiotica 1: 48-98.

Falkner, T. M. and J. DeLuce. 1989. Old age in Greek and Latin literature. Albany, NY: State University of New York Press.

Gamse, D. N. (Project Manager). 1984. Truth about aging: guidelines for accurate communications. Washington, DC: American Association of Retired Persons.

Gantz, W., H. M. Gartenber, and C. K. Rainbow. 1980. Approaching invisibility: the portrayal of the elderly in magazine advertisements. Journal of Communication 30: 56-60.

Gerlach, J. 1988. Adolescent literature: a misrepresentation of youth-aged relationships. Educational Gerontology 14: 183-191.

Gresham, M. 1976. The infantilization of the elderly. Nursing Forum 15: 196-210.

Kalish, N. 1979. The new ageism and the failure models: a polemic. The Gerontologist 19:398-402.

Kelley, C. E. 1983. Ageism in popular song: a rhetorical analysis of American popular song lyrics, 1964-1973. (Doctoral dissertation, University Oregon, 1982). Dissertation Abstracts International 43: 2830A.

Kent, K. E. M., and P. Shaw. 1980. Age in *Time*: a study of stereotyping. The Gerontologist 20:598-601.

Key, Mary Ritchie. 1975. Paralanguage and kinesics (Nonverbal communication). Metuchen, NJ: Scarecrow Press.

Kirch, M. S. 1979. Non-verbal communication across cultures. The Modern Language Journal 63: 416-423.

Knapp, Mark L. 1978. Nonverbal communication in interaction. 2nd ed. New York: Holt, Rinehart, and Winston.

Kubey, R. W. 1980. Television and aging: past, present, and future. The Gerontologist 20: 16-35.

Kvasnicka, B., B. Beymer, and R. M. Perloff. 1982. Portrayals of the elderly in magazine advertisements. Journalism Quarterly 54: 656-658.

Leitner, M. J. 1983. The representation of aging in pop/rock music of the 1960's and '70's. Activities, Adaptation and Aging 4: 49-53.

Levin, Jack, and Bill Levin. 1980. Ageism: prejudice and discrimination against the elderly. Belmont, CA: Wadsworth.

Lewis, M. I. 1973. Aging and mental health: positive psychological approaches. St. Louis: C. V. Mosby.

Lewis, Norman (ed.). 1964. The new Roget's thesaurus. New York: G. P. Putnam's Sons.

Makris, Mary and Stewart, Van Arthur. 2002. A course on women and aging in literature: Rationale, recommendations, and resources. Educational Gerontology 28(2): 121-138.

McKee, P., and H. Kaupinnen. 1987. The art of aging: a celebration of old age in western art. New York: Human Sciences Press.

Mathiason, G. 1977. It's time for fair play. Perspective on aging. 6: 17-22.

Meadows, R. E. 1986. The portrayal of older adults in basal reading textbooks of the 1960s and 1980s (Doctoral dissertation, University of Florida, 1986). Dissertation Abstracts International 47: 1673A-1674A.

Morris, William. 1979. The American heritage dictionary of the English language. Boston: Houghton Mifflin.

Murphy, J. E. 1980. The image of old age and the elderly in preindustrial America: a content analysis of selected eighteenth and nineteenth century magazines (Doctoral dissertation, Southern Illinois University, Carbondale, 1979). Dissertation Abstracts International 40: 4765A.

Nahemow, L., K. A. Mcluskey-Fawcett, and P. E. McGhee. 1986. Humor and aging. Orlando, FL: Academic Press.

Nuessel, Frank. 1982. The language of ageism. The Gerontologist 22:273-276.

____1984. Ageist language. Maledicta 8: 17-28.

____1988. Humanistic perspectives on aging. Paper presented at the 14th annual meeting of the Association for Gerontology in Higher Education, March 4, Chicago, IL.

____1991. Metaphor and cognition: a survey of recent publications. Journal of Literary Semantics 20: 37-52.

____1992a. The image of older adults in the media. Westport, CT: Greenwood Press.

____1992b. Philosophical and metaphorical aspects of language. Semiotica 89 (1-3): 117-128.

____1992c. The semiotics of ageism. University of Toronto Semiotic Circle Monograph Series, Vol. 10. Toronto: University of Toronto Semiotic Circle.

____1993. Metaphors and aging. Semiotica 97(1/2): 177-188.

____1994a. Metaphorical constructs, aging, and health care delivery. Signifying Behavior 1:201-213.

____1994b. Vico and contemporary views on language. In: M. Danesi and F. Nuessel, The imaginative basis of thought and culture: contemporary perspectives on Giambattista Vico, 85-97. Toronto: Canadian Scholars' Press.

____1995a. Future directions for research in the media depiction of older adults: a U. S. perspective. In C. Hummel, and C. J. Lalive D'Epinay, Images of aging in western societies, 287-319. Geneva, Switzerland: Centre for Interdisciplinary Gerontology, University of Geneva.

____1995b. Humanistic perspectives on aging: change and continuity in a humanities course one decade after its introduction. Paper presented at the annual meeting of the twenty-first annual meeting of the Association for Gerontology in Higher Education, February 25, Fort Worth, TX.

____1995c. Vico and current work in cognitive linguistics. In: M. Danesi, Giambattista Vico and Anglo-American science: philosophy and writing, 127-145. Berlin: Mouton de Gruyter.

____2002. Language about aging. In D. Ekerdt (ed.), Encyclopedia of Aging. 4 vols. New York: Macmillan Reference USA.

____1999. Ageism: Negative and Positive. 2nd ed. New York: Springer Publishing Co.

Palmore, Erdman. 1990. Ageism: negative and positive. New York: Springer.

Peterson, D. A, and E. L. Karnes. 1976. Older people in adolescent literature. The Gerontologist 16: 225-231.

Porter, L., L. M. Porter. 1984. Aging in literature. Troy, MI: International Book Publishers.

Ptacek, P. H., and E. K. Sander. 1966. Age recognition from voice. Journal of Speech and Hearing Research 9: 273-277.

Osgood, Charles E., George J. Suci, and Percy H. Tannenbaum. 1957. The measurement of meaning. Urbana: The University of Illinois Press.

Ramig, L. 1986. Aging speech: physiological and sociological aspects. Language and Communication 6: 25-34.

Range, J., and M. A. Vinovskis. 1981. Images of elderly in popular magazines: a content analysis of Littell's Living Age, 1845-1882. Social Science History 5: 123-170.

Ryan, E. and J. Capadano. 1978. Age perceptions and evaluative reactions toward adult speakers. Journal of Gerontology 33: 98-102.

Ryan, E. B., M. L. Hummert, and L. H. Boich. 1995. Communication predicaments of aging: patronizing behavior toward older adults. Journal of Aging and Social Psychology 14:144-166.

Seefeldt, C, A. Galper, K. Serock, and R. K. Jantz. 1978. The coming of age in children's literature. Childhood Education 54(3): 123-129.

Shipp, T. and H. Hollien. 1969. Perception of the aging male voice. Journal of Speech and Hearing Research 12: 703-710.

Smith. M. C. 1976. Portrayal of the elderly in prescription drug advertising: a pilot study. The Gerontologist 16: 329-334.

Smith, M. D. 1979. The portrayal of elders in magazine cartoons. The Gerontologist 19: 408-412.

Snider, James G., and Charles E. Osgood. 1969. Semantic differential technique: a sourcebook. Chicago: Aldine Publishing Company.

Stewart, Van Arthur. 1989. The use of popular print media visuals to dispel common negative attitudes toward older adults. In D. L. Gardner and M. C. Hoeckelman (eds.), Developing leadership in geriatric education: proceedings of the fourth annual summer institute, 155-157. Lexington, KY: Ohio Valley Appalachia Regional Geriatric Education Center.

Stewart, Arthur Van, and Frank Nuessel. 1996. Using humorous print materials in geriatric education programs to overcome negative public images of aging. Paper presented at the twenty-second annual meeting of the Association for Gerontology in Higher Education. Philadelphia, PA. March 3.

Stewart, Van Arthur. 1999. Research summary: Patronizing names and forms of address used with older people. Names 47(4): 401-409.

Stewart, Van Arthur and Cedeño, Aristófanes. 2001. A course on humanistic creativity in later life: Literature review, case histories, and recommendations. Educational Gerontology 27(8): 697-71§.

Stoddard, K. 1980. The image of women in American film, 1930-1980. (Doctoral dissertation, University of Maryland, College Park, 1980). Dissertation Abstracts International 41: 2663-A-2664-A.

Wang, C. 1988. LEAR'S Magazine "For the woman who wasn't born yesterday": a critical review. The Gerontologist 28: 600-601.

Wass, H., L. V. Hawkins, E. G. Kelly, C. R. Magners, and A. M. McMorrow. 1985. The elderly in the Sunday papers: 1963-1983. Educational Gerontology 11: 29-39.

Wisnieski, C. J., S. L. Star, and C. Hermann. 1979. Media guidelines for sexuality and aging. Mimeograph.

Woodward, K., and M. M. Schwartz. 1986. Memory and desire: aging-literature-psychoanalysis. Bloomington, IN: Indiana University Press.

Yahnke, R. E. 1996a. If youth only knew, if age only could: reflections on intergenerational relationships in feature-length films 1970-1995. Paper presented at the Twenty-second

Annual Meeting of the Association for Gerontology in Higher Education, March 2, 1996. Philadelphia, PA.

____1996b. Selected filmography: Intergenerational relationships in feature-length films 1970-1995. Paper presented at the Twenty-second Annual Meeting of the Association for Gerontology in Higher Education, March 2, 1996. Philadelphia, PA.

Yahnke, R. E., and R. M. Eastman. 1990. Aging in literature: a reader's guide. Chicago: American Library Association.

____1995. Literature and gerontology: a research guide. Westport, CT: Greenwood Press.

In: Advances in Communications and Media Research, Volume 4 ISBN 1-60021-189-5
Editor: William H. Miller, pp. 53-71

Chapter 4

PUBLIC AFFAIRS: COMPETITION AT THE PRODUCT LEVEL – THE NEWS

George Sylvie
University of Texas at Austin, USA

ABSTRACT

As American newspaper readership continues to decline, critics foresee adverse effects for the present-day public-affairs news product. Declining competition hints that newspapers will seek short-term profits over quality. But this chapter offers that newspapers have a strategic choice because of a more "dynamic" view of competition as posited by Bridges, Litman and Bridges (2002). The discussion uses the underlying assumptions inherent in the concept of competition and suggests that news and the newsroom are abstract, momentary phenomena that shift and move as fluidly as consumer behavior, prices, competition and management. In short, news is a multifaceted marvel and is discussed as such via the lenses of news production, the framing of competition, the economics of technology, the dynamics of costs, and the nuances of timing.

INTRODUCTION

America's newspaper industry faces an important junction. Most newspapers continue to lose circulation (Steinberg, 2004), a trend since 1987 (Compaine and Gomery, 2000). Swelling competition from other media for consumer attention, more demanding and market-savvy advertisers, and sluggish markets prevent significant growth (Compaine and Gomery, 2000). Meanwhile, critics say newspapers continued profit emphasis negatively affects the news product (e.g., Underwood, 1993; Squires, 1993) and journalists have begun to question the wisdom of that emphasis (e.g., Kaiser and Downie, 2002). Proposed solutions have varied, from the suggestion that the public had a right to the restoration of traditional journalism values (Kovach and Rosenstiel, 2001) to the call for a new business model (Picard, 1999).

One of the biggest concerns, however, is the impact of competition among newspapers. The last 20 years or so have seen a decline in multi-newspaper cities. With the increase in newspaper firm consolidation, fewer newspapers and publishers are contending for a dwindling audience (Compaine and Gomery, 2000). Although evidence suggests that larger chains and corporate papers tend to be as good as or better than non-corporate newspapers (Demers, 1996; Lacy and Fico, 1990), the fear remains that declining competition increases advertising costs and, hence, the likelihood that advertisers will seek better values for their money (Lacy and Martin, 2004).

A recent review of the literature reiterated that increased competition is good for newspaper content (Lacy and Martin, 2004). However, the evidence on declining competition hints that newspapers in such environments tend to seek short-term profits, cutting quality in the process. Doing so encourages competition from other media, including weekly newspapers and nearby dailies. Whether the news content in these other media approaches or exceeds the same level of quality as that of a metro daily constitutes a strategic choice and an uncertainty for the newspapers involved. For example, although we know that the circulations of competing metro and non-metro dailies are negatively related (e.g., Bridges, Litman and Bridges, 2002; Lacy and Sohn, 1990), we also know that competition isn't always adversarial, can vary, and may even be cooperative in nature - a more "dynamic" view of competition itself (Bridges, Litman and Bridges, 2002, p. 17).

As a result, a discussion of the impact of competition on the product – news, and especially public affairs content, requires a subtle, complex analysis. This chapter will examine news as product of competition and other factors, with the aim of describing its likely evolutionary development in the context of the more dynamic side of competition. We know, for instance, that there are organizational, social and cultural causes of media content (Shoemaker and Reese, 1991). Even within an economical context, a broad range of influences drive change in the newspaper industry (Picard, 2004).

But when looking at competition, the discussion must be informed by the underlying assumptions inherent in the concept. Competition, as discussed by scholars, tends to be viewed as a static, quantitative entity that can be packaged and plugged into any theoretical construct and measured and described via numbers. But competition, when it comes to both news and its source – the newsroom – is an abstract, momentary phenomenon that shifts and moves whenever a consumer consumes, a price fluctuates, a competitor performs, an editor edits or a publisher decides. In short, competition is a multifaceted phenomenon; so attempting to predict the future of newspaper content requires additional perspectives about the creation of news, the framing of competition, the economics of technology, the dynamics of costs, and the nuances of timing. The next stage of this chapter starts the process.

Creation of News

Defining News

To understand and predict whether news can change, we must closely examine what news is and how it happens. The definition of news is fluid, constantly changing with the situation. For example, if we define news as what's important, then hundreds of deaths at one time are important. But at times - especially amidst war - it may seem as if such deaths are commonplace or routine. Many journalists would say routines by nature and by definition do not constitute news, because routines are just that - routine, habitual, not easily noticed, the usual; in essence, they're not new. So if the *unusual* could be considered news, then when the city government gives away environmentally friendly low-flush toilets, it's news. Surely most city governments do not give away toilets as a matter of course. But if the city does so *every year,* then the action becomes routine and no longer news, although timing or qualifying criteria would be news. So just because an action is unusual, it also may be routine in a certain context, rendering it non-newsworthy.

Time elements pervade the news. For example, an event is said to be *timely* if it happens the day of or the day before publication or when it's due to happen in the immediate future. That's why the saying "That's old news" is apt: Anything old is not new, and certainly not news. So if a woman alleges that her supervisor sexually harassed her two years ago, it's generally not news unless associated with a more recent event. However, circumstances can render even "old news" newsworthy, as when a current Congressional candidate has his or her history examined by the public. Suppose the candidate had a drinking problem 10 years ago and has since abstained from alcohol. In such cases, other elements come into play. In this case, a Congressional race affects several thousand people and, as such, carries *impact.* That impact outweighs the timeliness of the action in helping to determine newsworthiness: Some will argue that a candidate's past behavior is an indicator or even a predictor of future behavior and the revelation of past indiscretions, such as alcoholism, is pertinent and, hence, news.

News Values

In other words, news is a dynamic concept, defined largely by journalists, and changing as situations change. It is defined in the context of other things, notably what people believe, say or do, as well as what they *don't* believe, *don't* say or refrain from doing. As a result, these and other elements - often called traditional *news values* - factor into the definition of news. In addition to the three mentioned above, there are others: for example,

- *human interest,* people or animals who do interesting things or who have interesting experiences;
- *prominence,* involving people, usually entertainers or politicians, widely known for their actions;
- *conflict,* when controversial actions or issues pit two or more people or groups or nonhuman elements such as tornados;

- *entertainment,* people, events or issues that amuse; and
- *trends,* patterns or shifts in behaviors of large groups of people (Rich, 2003).

These categories suggest that news just happens. But the above categories also reflect the input of audiences, who also create news. Journalists frequently get ideas from their readers or viewers. Public or "civic journalism" often features a forum, citizen panel or other public meeting in which ordinary citizens discuss their concerns about their community. Such concerns get converted to stories. Or editors assign stories based on what they believe the audience wants to read or view. For example, editors and others in positions of power in newspapers often want a reporter to do a particular story. Often this occurs because such managers believe the story would be interesting to audiences. They also may genuinely believe the story is "news." Suppose, for example, that gas prices escalate. An editor decides he will assign his reporters to get one-gallon gas cans, go to random gas stations and fill up the cans - all in the name of determining whether the station pumps accurately measure a gallon of gas. Such "consumer" news can be informative and entertaining, not to mention timely. But is it news? Many would say "no," but if the story appears in the newspaper the next day, some readers would be hard-pressed to argue that the story was not news. Such "news is what the editor says it is" fare isn't all that unusual.

In addition, many a reporter has gotten a story idea handed to him or her after the editor has had lunch with an influential citizen or group of individuals. These influentials often provide what are called “information subsidies,” prepared information regarding their own businesses or other interests. News also varies from publication to publication, mainly because each newspaper caters to an audience in a specific geographic region. Irrespective of geography, however, news depends on audiences because they expect news to be reported within the *social context* within which it occurs. And, the social context itself will vary, or change. When that happens, it's not unusual for news values to change. For example, the passage of time changes the audience’s tastes and how that audience views events, issues and people.

News Routines

But news often is created or manufactured from what has been called journalistic routines (Shoemaker and Reese, 1991). Such routines – the “habitual, ongoing, patterned procedures that are accepted as appropriate professional practice” (p. 13) – are used systematically by reporters and editors to ensure the newspaper will respond predictably. “Organizations must routinize work in order to control it” (p. 97); such processes take the reporter, for example, to the same sources, meetings and events.

Deadlines and writing formats constitute routines that result in constant similarity in presentation while helping the news organization to manage physical restraints. Deadlines, for instance, prompt the reporter to cease information-gathering, write a story, and adjust to the next story. Events occurring after the deadline have to wait until the next edition to get reported. Writing formats – such as the summary lead paragraph (also known as “the inverted pyramid” style) – force the reporter to think and organize information clearly and succinctly, with certain elements gaining priority in the story’s hierarchy. Anything too complex or trivial often doesn’t merit mention in that particular story.

New communication technology also influences how journalists produce news. Machines make possible speedy and widespread exposure of much more information about life and human behavior than ever thought likely. Computers, the Internet, television, film and other communication tools bring an abundance of data into our homes, in the process desensitizing the audience to certain topics; e.g., some scholars argue that repeated exposure to television cultivates – or teaches a common worldview – about certain issues, say, violence (Gerbner et al, 1994). Each new technology brings with it "new" kinds of content which, with widespread distribution, becomes "old" faster. For instance, the Internet has created an appetite for "instant" news, broadened the news cycle into "24/7" (all the time) and made even the smallest additional information in an ongoing crisis/event a "news" item. In addition, because the Internet allows anyone with a computer and a modem to "connect" to similar users and create a "webbed community," such an audience is able to choose and select information for its own interests. In the past, these people had no choice; they were forced to purchase a news medium or use it as the medium's producers saw fit. The news was what the news manufacturers, i.e., journalists, said it was. For Web users, news is now what they wish to view. In fact, some news organizations have begun to tailor their products or "push" certain variations of their products, such as local sports to such audiences (Toner, 1997).

So, news is fluid, dynamic and ever-changing, generally driven by the professional values of journalists and the working environment of the current newsroom. But news is also influenced in large measure by society in general. Traditional definitions of news exist, but those definitions are under constant bombardment because of the fact that news is a socially driven concept. As long as people change, so will news. So journalists not only help define news, but they must also keep abreast of change and adapt to it. Their jobs demand it. Research shows, for example, that the definition of news changed in the 20th century: Stories lengthened, included more analysis, expanded from specific to broader regions, placed more emphasis on time frames other than the present, and named fewer individuals and more groups, officials, and outside sources, especially affecting coverage of crimes, accidents, and employment. In short, journalists gradually added perspectives of social problems, interpretations, and themes to the news (Barnhurst and Mutz, 1997).

Competition as News

The Industry

When it comes to predicting the eventual news consequences of competition, current history suggests that change will not come without a struggle. A major struggle is the struggle over the demand for news. The international community, for example, provides a disappointing window into what the future may hold. In France, with national daily press sales figures continuing to drop, advertising revenue shares have shrunk, distribution costs have soared and the industry has become increasingly reliant on government subsidies. The resulting introduction of free dailies has not made a noticeable impact (Demoulins, 2004). In Germany, the situation parallels that of the United States: An increasing level of media selection; demographic changes in the population structure; and the loss of small, local newspapers; leading to severe advertising revenue cuts; and the disappearance of many inner city markets (Kopper, 2004). It would appear that we could expect a similar scenario, to some degree, in the United States.

But U. S. newspaper competition remains - in many ways - an unexplored or elusive phenomenon. For example, Lacy and Simon (1993) conceded that newspaper "competition" can simultaneously be both intense and non-existent; i.e., a newspaper can occupy a monopoly position and still behave as if it competes with another newspaper or media outlet. The impact of competition on the demand for news depends on what other media outlets deliver "news" and the degree to which these diverse outlets provide similar "news." Innate differences in media outlets make many of these versions flawed surrogates, suggesting that the situation that would produce the most perfect substitutes is direct competition between daily newspapers. But since direct competition is, for the most part, non-existent, then "umbrella competition" - competition between the newspaper and alternative forms such as weeklies, TV, radio and the Internet - may provide news options (Lacy and Simon, 1993).

For example, technology allowed smaller Japanese newspapers to cut costs and larger Japanese dailies to cover larger areas more cheaply. But although national and international coverage increased, local coverage suffered (Cho, 2002). Japanese newspaper managers apparently "decided to be good substitutes for, rather than complements to, national newspapers…." (p. 67). While inconsistent with what one would predict in the United States, where more local coverage would be expected, the study suggests that newspaper managers will consistently seek to expand their audiences, regardless of the competition. For example, the American newspaper industry acknowledges a renewed focus on younger, light readers (Nesbitt, 2003) and the need to learn how to market to demographically diverse populations (Wood, 2004), suggesting that the shape of competition constantly is in flux. As a result, a refined definition of competition – or at least a new way of measuring its fluidity – would significantly enhance the newspaper industry's ability to meet the challenge of creating appropriate "news" content.

The Journalistic Culture

Influential industry leaders advise (Davis, 2004) that to meet readers' needs, and thus ward off competition, the industry must not so much change news or "relevant content" but must "report and present news that resonates with readers" (p. 50). This challenge will require a newspaper to listen to its readers -- a significant change, considering journalists operate within a culture reluctant to change its professional approach to news (see Sylvie and Witherspoon, 2002). Changing content would require newspaper managers to engineer a cultural shift within the newsroom, no small feat given that cultural change requires managers to have a "temporal capability" – the ability to comprehend various seemingly opposite temporal conceptions about and dimensions of change, to discriminate among these conceptions and to use that knowledge, even if it means implementing one or more types of change behavior (Sylvie, 2003). A Readership Institute study (2000) emphasized changing culture.

A study of *The New York Times* shows that some editor/managers do have some temporal capabilities, assuming they know the history and the culture of their newspapers as well as does *Times* Publisher Arthur Sulzberger Jr. The *Times* as a unique case study suggests, however, that given the typical mid-level newsroom manager's five or fewer years of experience, most American newsrooms will have to wait for such skills and knowledge to develop. *The Times*' advantage as a "destination paper" means its managers probably have longer-than-typical managerial tenures, earning them adequate time to learn the newsroom culture (Sylvie, 2003). *The Times* case showed "a more enlightened approach" may be in store (Robertson, 2003). Although newspapers until recently historically neglected training newly promoted managers (Kees, 2002), current market changes may force the industry's hand. Still, profit-margin pressures have not reached the necessary threshold to prompt such training and until then, the journalistic notion of competition as some abstract notion – rather than a hardcore reality – likely will prevail (Sylvie and Witherspoon, 2002).

A more successful tack, however, may be starting to take hold in newsrooms. Many editors apparently plan to utilize their journalistic roots to prompt journalists to meet the competitive challenge, framing the notion of competition in journalistic terms. For example:

- An Idaho newspaper invited reader-parents to brainstorm with editors about important dates, such as report-card day, pertaining to schools. The effort is part of the Gannett newspaper group's "Real Life, Real News: Connecting with Readers' Lives" campaign structured to help Gannett newspapers reshape their approaches to news in order to be more "connected" with readers (Johnson, 2003).
- The Newspaper Association of America, in a quarterly newsletter to publishers and editors, showed how to reach diverse audiences through such methods as cross-departmental training, community "credibility roundtables" and special promotions that include the newsroom (Cottman, 2004).
- A Texas newsroom collaborated with the advertising department to create an online "virtual mall" in which teen girls recommend their male friends for a fashion makeover in the store of the sponsor. Photographs and a slideshow of the event appear on the newspaper's Web site and a story runs in the newspaper (Cottman, 2004).
- A South Carolina newspaper publishes a twice-weekly column featuring unknown, "ordinary people" who comment on life, books, films and inspirational personalities (Cottman, 2004).

Civic Journalism

Probably the most visible effort along these lines has been the civic journalism movement, which essentially gives readers and the public an increased voice in story ideas and in what constitutes news. As one active civic journalism editor-proponent put it: "We learned that our [the newspaper's] impact can be profound, if we really engage readers as citizens not just as an audience. We learned that we do better journalism if we're listening for more than just quotes. Our community sometimes can help us frame stories more intelligently" (Morgan, 2003). So, in a sense, editors have placed a "news" frame on competition and challenged journalists to cover more communities than before, not so much to change the way news is done as to seek "improved story-telling techniques," "greater accuracy," and "diversity." In essence, then, since trends in both market development and product demand are two key influences in competition (Wicks et al., 2004), newspaper managers simply have put a new face on them. Journalists – in their zest to "change the world" and to create stories with "impact" – in many cases have been willing accomplices to this facade.

Journalistic purists, however, would argue that such changes are not "news" in the traditional sense of journalism's institutional watchdog role in public affairs or as "the fourth branch of government," as it were. These purists would suggest that these changes cater more to what the audience *wants*, as opposed to what it *needs* – that no longer will we have a "marketplace of ideas" but only ideas driven by the market. The dilution of the journalistic filter or of journalistic interpretation, they would contend, yields not public affairs journalism but journalism by public opinion (see Kovach and Rosenstiel, 2001). In fact, they might say, this effort is not so much competition-based or competition-driven as it is profit-based and revenue-oriented. They would insist that civic practices -- despite their noble intent -- weaken journalism by making journalism an open advocate of an agenda instead of an impartial monitor. They would caution that the road to journalistic nirvana lies in giving the newsroom the autonomy and the resources with which to improve itself and not in quick, market-driven fixes that sacrifice trust – the newspaper's "brand equity" – at the expense of profit (Lacy and Martin, 2004; Meyer, 2004). Choosing the correct strategy often hinges on many things, not the least of which is technology.

BRANDING NEWS

Branding and Technology

In newspapers today, much is made of "extending the brand," a metaphor for increasing the recognition of the newspaper's identity or that with which it is most closely identified. For most newspapers "brand equity" is trust (e.g., see Kovach and Rosenstiel, 20011) and its brand is news or how it conveys that news. Technology plays a large role in news, from its collection to its packaging to its distribution. Many newspaper managers see technology as a market-changing tool, as well, and that change takes a variety of shapes.

For example, technology facilitates market entry for some while creating a barrier for others. The Internet has redefined markets for electronic newspapers. A newspaper in Houston that starts an online version will find competition for Houston's large Arab and Arab-American population from Middle Eastern newspapers as well. On the other end of the spectrum, that same technology allows the Houston paper to invade other metropolitan Texas markets and perhaps others outside of Texas, because the Houston paper can leverage its fairly exclusive coverage of the U.S. space program -- thanks to its proximity to the Johnson Space Center -- and thus appeal to readers who are somehow invested in space concerns (Chyi and Sylvie, 1998).

Blurring Markets

This blurring of markets doesn't necessarily stop at the newspaper's market boundaries, however; it may extend to the news itself. Because of the Internet's capability to deliver immediate feedback, a newspaper's online version becomes a method by which editors can gauge readers' interests on a variety of issues and topics. The journalistic danger is that readers' interests will drive editors' news selection, packaging or play and eventually outweigh journalistic judgment as to what's news. This temptation to take on the persona of an all-news cable TV station or a Web site – and let audience interests determine the subject of the moment (Downie and Kaiser, 2002) – is all too real for editors seeking to make their newspapers reader-friendly. This blurring is aided by the trend of newsroom technological convergence, in which journalists are asked to report and write for more than one medium: Doing essentially one story several times takes time away from reporting on other stories while trying to extend the newspaper's brand. The result may look impressive and save money, but the reader/viewer actually is denied the multiple perspectives that pre-convergent, competitive media normally would provide.

Another, more subtle side-effect of blurred, technology-driven markets lies in the uncertainty that the new market creates and, thus, the encouragement of conservative, *status quo* managerial approaches. When a newspaper enters a new market, unless experienced management accompanies the entry, there is an initial period of adjustment in which the newspaper attempts to clarify its goals relative to the new market. For example, although economic theory suggests that electronic newspapers should offer unique content to take advantage of the market-expanding power of the Internet, most newspaper companies tend to provide only slightly re-edited copy from the print edition for the online version, mostly as a cost-saving gesture. Newspaper online sites still tend to be fairly close facsimiles of the print edition. In fact, in the early days of electronic newspapers, some online newspaper executives showed difficulty defining their market and primary competitor (Chyi and Sylvie, 2000).

Readership

A study of context-leading U.S. dailies and their Web sites showed none apparently used the Internet's potential to add more context to their environmental articles beyond additional Web links (Randazzo and Greer, 2003). And while perhaps a victory for the traditional-news formula, this easy, repetitive approach eventually is counterproductive for attracting new readers. Such replication suggests that while newspaper online sites offer the same product as their print companions -- thus extending the brand -- that strategy does not necessarily appeal to or recruit non-newspaper readers. The readers of online editions tend to also be readers of the print edition and, given the same price and the same content, Web users have overwhelmingly favored the print format (Chyi and Lasorsa, 1999). Today, most newspaper Web site readership continues to be a duplication of the print audience; in a typical week, 54 percent of general online users read only the print product, 20 percent read both the paper and the Web site, and 6 percent read only the Web site – figures unchanged from 2002 figures. The news gets worse: Some 40 percent of all Internet users who went online in 2004 for local news use newspaper Web sites, down from 62 percent in 2002. Online users are turning more toward search-engine sites, such as Google and Yahoo! for online content (Coats, 2004).

Oddly enough, this trend surfaces despite the fact that the evolving nature of news has forced journalists to be sensitive to audience tastes, prompting the traditional public affairs news topics to fall from favor. A recent content analysis of 16 newspapers' front pages showed government news in 2003 made up a fairly smaller percentage of front-page articles than 20 years before. Foreign affairs made up a smaller number of the articles on the front page than in the Cold War-influenced 1970s, but about the same as 1997. The number of articles about crime kept falling while the number of lifestyle articles continued to climb (Project for Excellence, 2004). So the Internet, while slightly widening the audience for traditional news, doesn't necessarily guarantee that news will have lasting appeal to the new audience. The phenomenon of convergence acts similarly, expanding quantity but not necessarily enhancing the quality of news. In fact, because quality in news tends to mean more instead of fewer words, the Internet's inherent nature of immediate feedback and graphic interfacing more likely will discourage quality storytelling -- not surprising, since technological advances alter habits and tasks. As a result, online newspapers typically use consumer services on their sites. Such services usually allow the reader to insert some form of input – a value, symbol or word – with a resulting bit of information from a database – e.g., forums, email, and site searches. Few newspapers use plug-in based technologies such as chat rooms, other languages, updated polls, audio and video (Dibean and Garrison, 2001). Translation: Newspapers tend to not take full advantage of the Internet's capabilities. Although gaining more readers is usually the main reason for publishers to put their papers on the Web (Peng, Tham and Xiaoming, 1999), the sites offer mostly traditional text and low-interactive features that, again, expand the brand but don't always bring in new readers. So the outlook for traditionally written news -- in an online format, at least -- as a magnet for new readers looks dim.

Opportunity Costs

The "Cost" of News

When scholars think of the cost of news, they typically look toward the newsroom and refer to the price of labor and various production and structural elements (see Lacy and Simon, 1993; Picard, 1993). But there are other costs, such as those associated with audience gratification, as when controversial material or opinions upset readers and cause them to stop reading. For newspaper readers, newspapers are sources of gratification; thus, so is news. So competition for readers means newspapers have to offer news that will prompt readers to buy the newspaper or to allocate time and attention to it – the medium's opportunity costs (Nicholson, 1979).

Inherent in this potential transaction and the concept of attention is the assumption that the ones who produce the news have a solid grasp on what those who consume the news expect and want. A closer look, however, tells us that even though the topics of news have changed over the years (Project for Excellence, 2004), there is no conclusive proof that journalists are in step with their audiences. For example, Stone and Boudreau (1995) found newspaper readers' content preferences were stable over a decade – suggesting a hopeful omen for the newspaper industry and a tacit endorsement of the notion of gradual modification of news without alienating core readers. But signs indicate that editors often err in foreseeing reader interest. For example, a survey of Ohio newspaper editors and the general Ohio public showed editors don't know what their readers want; outside of local health and school news, the Ohio editors incorrectly predicted interest in seven of eight categories (Bernt et al., 2000).

Reader Interest

But "interest" alone doesn't tell the whole story, as the latest industry numbers illustrate. Readers' tastes vary. For example, although readers' news and feature interests have remained fairly stable, certain types of interests are more powerful than others in driving frequent newspaper readership: Local, state, national and international news, opinion and analysis, and business news are powerful drivers (see Table 1). Frequency of newspaper readership is not strongly tied to entertainment or feature interests. As to how newspaper readers would like to see news media content organized, 70 percent or more list timeliness, credibility, being interesting, depth of coverage and ease of use as very important. Personally appealing values – e.g., cares about people like you, reflects your personal beliefs, provides positive news, gives you things to talk about, is enjoyable and entertaining – don't rate nearly as high. And while satisfaction with newspaper content also remains stable, readers want improvement in news that helps communities solve problems and advances in opinion and analysis. The opportunity for increased market share from increased local news coverage is that much stronger, considering that newspapers dominate other media when people rate their satisfaction in local news (62 percent, with TV second at 48 percent). But in other areas – weather, world and national news, state news, environmental news, and business news – newspapers are in intense competition with other media (Newspaper Association of America, 2000).

Table 1. Interest in Content Areas (in percent)

Type of Content	Frequency of Weekday Readership		
	Total	Regular Readers (4-5 issues)	Occasional Readers (1-3 issues)
Local News (Avg).	64	71	62
Local Government and Political News	54	64	46
State News	61	69	60
World and National News	69	76	70
Opinion and Analysis	37	43	33
Local Business News	47	53	46
National Business News	43	50	39

Source: NAA, 2000

The industry knows, then, that readers understand what newspapers have to offer as well as their differences in coverage. Broadening coverage, changing how news is written and more effective promotion of news helps readership (Readership Institute, 2001). So "interest" can be heightened – and "costs" lessened – if newspapers follow the blueprint set out for them (pp. 14-15). The more time newspapers spend to improve key content areas, the more likely they are to prompt readers to spend with the newspaper and to increase the number of readers (pp. 8-10). Currently, Americans who read a newspaper spend an average of slightly more than 30 minutes per day (Newspaper Association of America, 2000). The easing of such "opportunity costs" where it concerns news obviously is a strong priority of the industry. But as our examination of news' origin, competitive frame, technological extensions, and audience-related costs demonstrate, the news is a time-sensitive, labor-intensive puzzle of competitive elements that make up these opportunity costs. Assessment of the most direct, competitive influence – management – promises to put the picture into perspective.

CHANGING NEWS ROUTINES

The Role of Management

The previous discussion has left little doubt that if news is to remain essential to consumers, changes must occur. Newspapers with the requisite "constructive" cultures tend also to have higher readership (Readership Institute, 2000). Competition is positively related to the number of full-time newsroom employees, which in turn is related to quality (Lacy and Blanchard, 2003). Since newsroom managers decide staffing levels and because most U.S. newspaper markets do not have intense newspaper competition, editors and their sub-editors become important arbiters regarding news and its relationship to readers. And how these editors manage their subordinates holds an important key as to whether traditional news formats – whether for traditional or nontraditional news – can be changed to increase readership or whether news will gradually disappear from the consumer menu.

The need to change has been chronicled at length elsewhere, as has the role of managerial timing in changing these newsroom routines (Sylvie and Witherspoon, 2002; Sylvie, 2003). Suffice it to say that the self-perpetuation of the traditional newsroom culture and the lack of newspaper managerial expertise with the nuances of timing do not bode well for the future of news. The various environmental influences on newspapers naturally impede change. For example, technological convergence, the Internet's growing influence, and continuing government regulation of cross-media ownership have forced newspapers into an uncertain, "defensive" posture (Singer, Tharp and Haruta, 1999; Wicks et al., 2003; Sylvie and Witherspoon, 2002). Coupled with organizational differences within newspapers, slow and limited change characterizes the industry. Newspapers also feel increasing pressure to reflect American society's transformation into a more diverse body. Statistical differences between newsroom sub-cultures present an ample hurdle to hiring and maintaining diverse cultures, as well as preventing them from learning from each other and thus creating news of interest to new, potential readers (Lewis, 1995; Gentry, 1997; Gade, 1999; TeBoul, 1999; Gade, 2000; Readership Institute, 2000). But journalists – using occupationally and culturally based notions of how to do journalism – typically resist change, viewing it as troublesome or unproductive. After a short period of anticipation, they misunderstand editors' message of change (Neuzil et al., 1999; Sylvie and Witherspoon, 2002; Gade and Perry, 2003).

Management and Timing

Hence the need for timing, and not simply as the already accepted views of time as a structural journalistic element that affects routines of newsgathering and editing, production deadlines and the very definition of news. Other timing dimensions abound. For example, management scholars have discovered that the mechanisms that groups use to *pace* their work may critically influence their collective incentive to change. Timing also is a function of context, or *plot*, and individuals behave in agreement with the meaning and requirements of the moment: There's clock time – employees' work routines, inner time – their psychological comfort, and social time – the quality of their relationships. For example, a newspaper adopting a new news format should give its reporters ample time, training, and motivation to learn the new system. Other aspects of time include sequencing – the order in which events unfold, timing – the moment an event happens or is planned to happen in a sequence of related events, pacing – the rate or rhythm at which change occurs, and a combination of the three (Huy, 2001; Sylvie and Witherspoon, 2002).

Yet another aspect concerns timing norms: shared, expected patterns of paced activity, such as newspaper deadlines, that regulate managerial activities. Some norms, such as coffee breaks, surface via social contact, while editors impose others, for example, news budget meetings. Others, such as news teams, come about via coordination or through management formalization, e.g., performance appraisals. Clearly, timing norms pervade a newspaper – and any organization's – management activities, programs and events (Ancona et al., 2001). If newspapers are to successfully and managerially "time" the needed change in news, editors will have to amend the pace of newsgathering – a practice termed "entrainment." Scholars suggest that successful entrainment must be proactive as well (Ancona and Chong, 1996; Rosenbloom, 2000). In essence, then, the management of newsgathering must take on a more thoughtful approach.

Timing of Competition

Underlying the timing of news, however, is the timing of competition. In this context, timing suggests a kind of structure that "creates" "opportunities." Such "temporal structuring" (Orlikowski and Yates, 2002) shapes a rhythm and form, but depends on viewpoint, as shaped by moment of observation. Temporal structures often are routine, taken for granted and can overlap (pp. 686-687). One such structure is the business lifecycle, the stage of which can determine when a new competitor arises or when a veteran competitor exits (Caves, 1998). According to this cycle, declining demand for the product pressures production to decrease or even stop in order to preserve profits (Ghemawat and Nalebuff, 1985). Survival of firms with unequal market segments but similar costs then depends on those firms' market rank. Sufficiently strong scale economies can give large firms enough cost advantages to force out smaller and less efficient firms (Fudenberg and Tirole, 1986). Of the cycle stages of introduction, growth, maturity and decline, the newspaper industry as a whole is thought to be in the latter stages of maturity (Picard, 1999). A study of the exit rates in the daily newspaper market in The Netherlands from 1950 to 1996 found the decisions to exit are time-dependent, i.e., in concert with the life-cycle idea (van Kranenburg, Palm, and Pfann, 2001).

News and Temporal Structure

To manage the temporal structure of news, a time-sensitive newspaper editor would not have to be aware of the industry lifecycle itself in order to change the content and format of traditional news. Instead, a wise editor knows that every journalistic format has its own lifecycle, and that the time will come for each news format to change, or die. Market influences – in the form of increased demands on readers' time and the advent of televised news – on the inverted pyramid or summary-style first paragraph news structure begot the "soft lead," which then begot the "narrative," or storytelling-oriented structures that feature a "human element" (see Rich, 2003). Taking advantage of journalists' natural desire to have impact on readers, enterprising editors must begin to see each story as an opportunity for change. The difficulty, however, lies in the fact that the typical editor often is too constrained and consumed by each day's routine to clearly see when a format's decline looms on the horizon; this is especially true at smaller newspapers (e.g., Vandini, 2000). So not only is time the editor's friend, it also is the editor's enemy. Were it not for larger, trend-setting newspapers such as *The New York Times, USA TODAY, The Washington Post*, or, more typically, the larger newspapers in the various, respective newspaper conglomerates, innovations would not make their way to smaller, more typical-sized newspapers. This "trickle-down effect" resembles the diffusion process (Rogers, 1983) routinizing stage – where a technology or idea is firmly in place, is part of work, and often becomes the key stage of change. An employee-idea-technology interaction affects editors' perceptions of the technology/idea utility and flexibility. The newsroom then gains knowledge causing it to revise its approach, which then will lead it to change the preferred level of use of the technology-idea. This "education device" self-corrects to influence newsroom capabilities and growth risks. For example, one newsroom changed its structure to topic-team reporting when shown that some topics - such as science, health and medicine - were inadequately covered. As a result, health and science received a much greater share of news space. But there also were policy changes in coverage and display: in essence, the newsroom adopted a new set of news values (Russial, 1997). So news will not change unless editors initiate it, follow through, listen to reporters' and market feedback, and adjust accordingly. To start, editors themselves must be attuned to competitive forces or, at the least, to their peers at larger, better newspapers.

CONCLUSIONS

News is Solidly Fluid

This analysis obviously does not sound the death knell for news and actually focuses on presentation as much as content change. If anything, it assumes that newspapers will change as needed and public affairs news will survive for the near future, provided the definition of news remains dynamic, with fairly consistent input from editors, journalists, readers and the market in general.

This view doesn't question the prevailing wisdom that lack of competition may signal the death of newspaper quality; most evidence points in that direction (Lacy and Martin, 2004). However, how management views competition is all-important in the long-range scheme of things. Lack of newspapers doesn't necessarily translate exactly to "lack of competition." Journalists' much-revered notion of autonomy – as embodied in the professional approach – in its purest sense assumes that the journalist will strive for self-improvement and continued service to the reader. It also is assumed that the editor will continue to serve the needs of the organization in seeing that the journalist has – on most days – every resource with which to accomplish the professional ideal.

Much will hinge on the abilities of editors, usually as former reporters, to speak on behalf of their staffs to their publishers and to aid those publishers in continuing to promote to corporate ownership the idea that news is the strength of the newspaper. The maturity stage of the newspaper lifecycle doesn't connote staleness. The very word "maturity," while it suggests a kind of ripeness, also intimates the supposed prime of life. So instead of being threatened by innovations such as the Internet and convergence, newspapers and news instead can see these innovations as opportunities to not only "extend the brand," but as chances to transform – in terms of delivery, "24/7," *and* in terms of substance.

Online News

In the case of the online newspaper, for example, an early analysis (Chyi and Sylvie, 1998) of the electronic newspaper market suggested it is four-fold, with different types of competition at varying intensity levels. The analysis encouraged product differentiation as a key to market success and power – a process implying and requiring multiple market definitions (pp. 16-17). For example, "local news" in the printed newspaper traditionally is largely defined as what happens of import and of interest within the city proper, be it police news, school news, city hall news, and the like. But for the online newspaper "local news" can cover a wide range; for instance, police news to the reader in the suburb may be news about the influx of drugs from the city, while police news to the reader 200 miles away may be a story about the overall crime trend in the city. "Local news" takes on a different meaning, depending on the reader's geographic locale (Johnson, 2004). In an online context, then, an online newspaper market casts doubt about the usefulness of normal market analysis models, in much the same fashion as Bridges et al. (2002) did in suggesting "a variety and fluidity of market roles and metropolitan geographic relations" (p. 17).

The Future

When newspapers will reach the "decline" stage of their lifecycle and when news will no longer matter will become apparent soon enough: In time, journalists may lose sway over the newsroom, technology may advance to the point of making the news form obsolete, or competition from other media or from other kinds of newspapers may so intensify as to create new formats. In any event, it certainly will be news.

References

Albert, S. (1995). Towards a theory of timing: An archival study of timing decisions in the Persian Gulf War. *Research in Organizational Behavior, 17*: 1-70.

Ancona, D., and Chong, C.-L. (1996). Entrainment: Pace, cycle, and rhythm in organizational behavior. *Research in Organizational Behavior 18*: 251-284.

Ancona, D. G.; Goodman, P. S.; Lawrence, B. S.; and Tushman, M. L. (2001). Time: A new research lens. *Academy of Management Review 26*(4): 645-663.

Barnhurst, K., and Mutz, D. (1997). American journalism and the decline in event-centered reporting. *Journal of Communication 47*(4): 27-53.

Bernt, J. P.; Fee, F. E.; Gifford, J.; and Stempel, G. H. III (2000). How well can editors predict reader interest in news? *Newspaper Research Journal 21*(2): 2-10.

Bridges, J. A.; Litman, B. R.; and Bridges, L. W. (2002). Rosse's model revisited: Moving to concentric circles to explain newspaper competition. *The Journal of Media Economics 15*(1): 3-19.

Caves, R. E. (1998), Industrial organization and new findings on the turnover and mobility of firms. *Journal of Economic Literature (36)*: 1947-1982.

Cho, H. (2002). The effect of competition among Japanese local and national dailies on news coverage in local newspapers. *Keio Communication Review 24*: 66-70.

Chyi, H. I., and Lasorsa, D. (1999). Access, use, and preferences for online newspapers. *Newspaper Research Journal, 20*(4): 2-13.

Chyi, H. I., and Sylvie, G. (1998). Competing with whom? Where? And how? A structural analysis of the electronic newspaper market. *Journal of Media Economics 11*(2): 1-18.

Chyi, H. I., and Sylvie, G. (2000). Online newspapers in the U.S. -- perceptions of markets, products, revenues and competition. *The International Journal on Media Management 2*(2): 69-77.

Coats, R. (2004). Power users 2004: Newspapers' online audience in a broadband world. Vienna, Va.: Newspaper Association of America.

Compaine, B. M., and Gomery, D. (2000). *Who owns the media: Competition and concentration in the mass media industry*. Mahwah, NJ: Erlbaum.

Cottman, M. H. (2004, Spring). Connecting to new readers. *Fusion*: 1-6.

Davis, A. B. (2004, January). A presciption for surviving change. *presstime 26*(1): 50.

Demers, D. (1996). Corporate newspaper structure, editorial page vigo and social change. *Journalism and Mass Communication Quarterly 73*(4): 857-877.

Demoulins, N. D. (2004). Recent evolution of the French newspaper industry: Decreasing advertising and the appearance of free newspapers. In Picard, R.G. (ed.), *Strategic responses to media market changes*. Jönkoping, Sweden: Jönkoping International Business School, pp. 97-103.

Dibean, W., and Garrison, B. (2001). How six online newspapers use Web technologies. *Newspaper Research Journal, 22*(2): 79-93.

Downie, L. Jr., and Kaiser, R. G. (2002). *The news about the news: American journalism in peril.* New York: A.A. Knopf.

Fudenberg, D., and Tirole, J. (1986). A theory of exit in duopoly, *Econometrica (54):* 943-960.

Gade, P. (2000). The cultural transformation of U.S. newspapers: A comparison of management and rank-and-file attitudes towards a conceptual model of organizational development. Paper presented at the annual meeting of the Association for Education in Journalism and Mass Communication. Phoenix, AZ.

Gade, P. J., and Perry, E. L. (2003). Changing the newsroom culture: A four-year case study or organizational development at the *St. Louis Post-Dispatch*. *Journalism and Mass Communication Quarterly 80*(2): 327-347.

Gentry, J. K. (1997). Keys to success: Change survey results. In ASNE (Eds.), *Change: Living it, embracing it, measuring it.* Reston, VA: American Society of Newspaper Editors Change Committee.

Gerbner, G., L. Gross, M. Morgan, and N. Signorielli (1994). Growing up with television: The cultivation perspective. In J. Bryant and D. Zillman (eds.), *Perspectives on media effects*, pp. 17-40. Hillsdale, NJ: Erlbaum.

Ghemawat, P., and B. Nalebuff (1985). Exit. *Rand Journal of Economics (16):* 184-194.

Huy, Q. N. (2001). Time, temporal capability, and planned change. *Academy of Management Review 26*(4): 601-623.

Johnson, T. D. (2003, December). Hurricane can't stop "Real News." *presstime 25*(12): 22.

Johnson, T. D. (2003, September). Local news: How local can you go. *presstime 25*(9). Retrieved Oct. 4, 2004, from http://www.naa.org/preview.cfm?AID=5153.

Kees, B. (2002). *Newsroom training: Where's the investment? A study for the Council of Presidents of National Journalism Organizations.* Miami: The Knight Foundation.

Kovach, B., and Rosenstiel, T. (2001). *The elements of journalism: What newspeople should know and the public should expect.* New York: Three Rivers Press.

Kopper, G. G. (2004). The German newspaper industry: A case study in changing media markets and strategic options for the European newspaper industry. In R. G. Picard (Ed.). *Strategic responses to media market changes.* Jönkoping, Sweden: Jönkoping International Business School, pp. 105-119.

Lacy, S., and Blanchard, A. (2003). The impact of public ownership, profits, and competition on number of newsroom employees and starting salaries in mid-sized daily newspapers. *Journalism and Mass Communication Quarterly 80*(4): 949-968.

Lacy, S., and Fico, F. (1990). Newspaper quality and ownership: Rating the groups. *Newspaper Research Journal 11*(2): 42-56.

Lacy, S., and Martin, H. (2004). Competition, circulation and advertising. *Newspaper Research Journal 25*(1): 18-39.

Lacy, S., and Simon, T. F. (1993). *The economics and regulation of United States newspapers.* Norwood, NJ: Ablex.

Lacy, S., and Sohn, A. B. (1990). Correlations of newspaper content with circulation in the suburbs: A comparative case study. *Journalism Quarterly 67*(4): 785-793.

Lewis, R. L. (1995). Transformed work organization in newspaper firms (Doctoral dissertation, University of North Carolina at Chapel Hill, 1995). *Dissertation Abstracts International, 51,* 417.

Meyer, P. (2004). The influence model and newspaper business. *Newspaper Research Journal 25*(1): 66-83

Morgan, A. N. (2003, September-October). Connecting with the community. *The American Editor 78*(5): 36-40.

Nesbitt, M. (2003, December). Targeting growth. *presstime 25*(12): 24-25.

Neuzil, M.; Hansen, K.; and Ward, J. (1999). Twin Cities journalists' assessment of topic teams. *Newspaper Research Journal, 20*(1): 2-16.

Newspaper Association of America (2000). Leveraging newspaper assets: A study of changing American media usage habits. Vienna, Va.: Author.

Nicholson, W. (1979). *Intermediate microeconomics and its application* (2nd ed.). Hinsdale, IL: Dryden.

Orlikowski, W. J., and Yates, J. (2002). It's about time: Temporal structuring in organizations. *Organization Science 13*(2): 684-700.

Overholser, G. (2004). Good journalism and business: An industry perspective. *Newspaper Research Journal 25*(1): 8-17.

Peng, F. Y.; Tham, N. I.; and Xiaoming, H. (1999). Trends in online newspapers: A look at the U.S. Web. *Newspaper Reseach Journal, 20*(2): 52-63.

Picard, R. G. (1993). Economics of the daily newspaper industry. In A. Alexander, J. Owers, and R. Carveth (Eds.). *Media economics: Theory and practice.* Hillsdale, NJ: Erlbaum Associates.

Picard, R. G. (1999, 2nd Quarter). Implications of the changing business model of the newspaper industry. *International Newspaper Financial Executives Quarterly Journal 5*(3): 2-5.

Picard, R. G. (2004). Environmental and market changes driving strategic planning in media firms. In R. G. Picard (Ed.). *Strategic responses to media market changes.* Jönkoping, Sweden: Jönkoping International Business School, pp. 1-17.

Project for Excellence in Journalism, The (2004). The state of the news media 2004. Washington, D.C.: Author. Retrieved Oct. 4, 2004, from http://www.stateofthenewsmedia.org/index.asp

Randazzo, R., and Greer, J. (2003). Newspapers' Web sites add little to print version of environmental news. *Newspaper Research Journal, 24*(2): 120-124.

Readership Institute – Media Management Center at Northwestern University (2000). *Culture report: A profile of the impact newspapers and their performance.* Evanston, IL: Author.

Readership Institute – Media Management Center at Northwestern University (2001). *Newspaper Content: What makes readers more satisfied.* Evanston, IL: Author.

Rich, C. (2003). *Writing and reporting news.* Belmont, CA: Wadsworth.

Robertson, L. (2003, August/September). Down with top-down. *American Journalism Review25*: 36-44.

Rogers, E. (1983). *Diffusion of innovation (3rd ed.).* New York: Free Press.

Rosenbloom, R. (2000). Leadership, capabilities, and technological change. *Strategic Management Journal, 21*: 1083-1103.

Russial, J. T. (1997). Topic-team performance: A content study. *Newspaper Research Journal, 18*(1-2), 126-144.

Shoemaker, P. J., and Reese, S. D. (1991). *Mediating the message: Theories of influences on mass media content.* New York: Longman.

Singer, J. B.; Tharp, M. P.; and Haruta, A. (1999). Online staffers: Superstars or second-class citizens? *Newspaper Research Journal, 20*(3): 29-47.

Squires, J. D. (1993). *Read all about it: The corporate takeover of America's newspapers.* New York: Times Books.

Steinberg, J. (2004, May 4). Newspaper circulation continues overall decline. *The New York Times.* C4.

Stone, G. C., and Boudreau, T. (1995). 1985, 1994: Comparison of reader content preferences. *Newspaper Research Journal 16*(4): 13-28.

Sylvie, G., and Witherspoon, P. D., (2002). *Time, change, and the American newspaper.* Mahwah, NJ: Erlbaum.

Sylvie, G. (2003). A lesson from *The New York Times*: Cultural change and the role of managerial timing. *The International Journal on Media Management 5*(4): 295-305.

Teboul, J. C. B. (1999). Racial/ethnic "encounter" in the workplace: Uncertainty, information-seeking, and learning patterns among racial/ethnic majority and minority new hires. *The Howard Journal of Communications 10*(2): 97-121.

Toner, M. (1997, March). The push is on. *presstime*, pp. 41-46

Tuchman, G. (1973). Making news by doing work: Routinizing the unexpected. *American Journal of Sociology 79*: 110-131.

Underwood, D. (1993). *When MBAs rule the newsroom.* New York: Columbia University Press.

van Kranenburg, H. L; Palm, F. C.; and Pfann, G. A. (2001, August 11). Survival in a concentrating industry: The case of daily newspapers in The Netherlands. Paper presented to The 15th Nordic Conference on Media and Communication Research. Reykjavik, Iceland. Unpublished paper.

Vandini, C. (2000). It's not about the money: Small-town newspapering in Texas. Austin, TX: Eakin Press.

Wicks, J. L.; Sylvie, G.; Hollifield, C. A.; Lacy, S.; and Sohn, A. B. (2004). Media management: A casebook approach (3rd. ed.). Mahwah, NJ: Erlbaum.

Wood, C. (2004, Spring). Ethnic markets key to new revenue. Fusion: The business of diversity in the newspaper industry: p. 7.

In: Advances in Communications and Media Research, Volume 4 ISBN 1-60021-189-5
Editor: William H. Miller, pp. 73-101

Chapter 5

REGULATION OF BROADCAST INDECENCY: BACKGROUND AND LEGAL ANALYSIS*

Angie A. Welborn and Henry Cohen

ABSTRACT

Two prominent television events in the past two years placed increased attention on the FCC and its indecency regulations. The airing of the 2003 Golden Globe Awards and the subsequent ruling by the FCC's Enforcement Bureau, coupled with the controversy surrounding the 2004 Super Bowl halftime show, brought broadcast indecency to the forefront of the congressional agenda during the 108th Congress. Several bills were introduced, but not enacted, in the 108th Congress to increase the penalties imposed for broadcast indecency and to prohibit the broadcast of certain words and phrases in any grammatical form. Similar legislation aimed at increasing penalties has been introduced in the 109th Congress (H.R. 310, S. 193, S. 616). Other legislation addressing specific types of programming has also been introduced (H.R. 1420 and H.R. 1440). This report provides background on the two events in question, discusses the legal evolution of the FCC's indecency regulations, and provides an overview of how the current regulations have been applied and pending legislation. The final section of the report considers whether prohibiting the broadcast of "indecent" words regardless of context would violate the First Amendment.

INTRODUCTION

Two prominent television events in the past two years placed increased attention on the Federal Communications Commission (FCC) and its broadcast indecency regulations.[1] The airing of the 2003 Golden Globe Awards and the subsequent ruling of the FCC's Enforcement Bureau, coupled with the controversy surrounding the 2004 Super Bowl half-

---

* Excerpted from CRS Report RL 32222 dated December 2, 2005.

time show, have brought broadcast indecency to the forefront of the congressional agenda. During the 108th Congress, several bills were introduced to increase the penalties imposed for broadcast indecency and prohibit the broadcast of certain words and phrases though none of the legislation was enacted. In addition, both the House and Senate held hearings on broadcast indecency. Similar legislation has been introduced in the 109th Congress. This report provides background on the two events in question, discusses the legal evolution of the FCC's indecency regulations, and provides an overview of how the current regulations have been applied and pending legislation. The final section of the report considers whether prohibiting the broadcast of "indecent" words regardless of context would violate the First Amendment.[2]

BACKGROUND

On January 19, 2003, a number of broadcast television stations in various parts of the country aired the Golden Globe Awards. During the awards, the performer Bono, in response to winning an award, uttered the phrase "this is really, really f[***]ing brilliant."[3] In response to this utterance, the FCC received over 230 complaints alleging that the program was obscene or indecent, and requesting that the Commission impose sanctions on the licensees for the broadcast of the material in question.[4]

The Enforcement Bureau of the FCC issued a Memorandum Opinion and Order on October 3, 2003, denying the complaints and finding that the broadcast of the Golden Globe Awards including Bono's utterance did not violate federal restrictions regarding the broadcast of obscene and indecent material.[5] The Bureau dismissed the complaints primarily because the language in question did not describe or depict sexual or excretory activities or organs. The Bureau noted that while "the word 'f[***]ing' may be crude and offensive," it "did not describe sexual or excretory organs or activities. Rather, the performer used the word 'f[***]ing' as an adjective or expletive to emphasize an exclamation."[6] The Bureau added that in similar circumstances it "found that offensive language used as an insult rather than as a description of sexual or excretory activity or organs is not within the scope of the Commission's prohibition on indecent program content."[7]

The decision of the Enforcement Bureau was met with opposition from a number of organizations and Members of Congress, and an appeal was filed for review by the full Commission. FCC Chairman Michael Powell asked the full Commission to overturn the Enforcement Bureau's ruling.[8]

On March 18, 2004, the full Commission issued a *Memorandum Opinion and Order* granting the application for review and reversing the Enforcement Bureau's earlier opinion.[9] The Commission found that the broadcasts of the Golden Globe Awards violated 18 U.S.C. 1464, but declined to impose a forfeiture on the broadcast licensees because the Order reverses Commission precedent regarding the broadcast of the "F-word." This decision is discussed in detail below.

On February 1, 2004, CBS aired Super Bowl XXXVIII, with a half-time show produced by the MTV network. The show included performers singing and dancing provocatively, and ended with the exposure of the breast of one female performer. The network received numerous complaints regarding the half-time performance and FCC Chairman Michael Powell initiated a formal investigation into the incident.[10]

On September 22, 2004, the FCC released a *Notice of Apparent Liability for Forfeiture* finding that the airing of the Super Bowl halftime show "apparently violate[d] the federal restrictions regarding the broadcast of indecent material."[11] The *NAL* imposes a forfeiture in the aggregate amount of $550,000 on Viacom Inc., the licensee or ultimate parent of the licensees with regard to whom the complaint was filed.[12] This decision is discussed in detail below.

EVOLUTION OF THE FCC'S INDECENCY REGULATIONS

Title 18 of the United States Code makes it unlawful to utter "any obscene, indecent, or profane language by means of radio communication."[13] Violators of this provision are subject to fines or imprisonment of up to two years. The Federal Communications Commission has the authority to enforce this provision by forfeiture or revocation of license.[14] The Commission's authority to regulate material that is indecent, but not obscene, was upheld by the Supreme Court in *Federal Communications Commission v. Pacifica Foundation.*[15] In *Pacifica,* the Supreme Court affirmed the Commission's order regarding the airing of comedian George Carlin's "Filthy Words" monologue.[16] In that order, the Commission determined that the airing of the monologue, which contained certain words that "depicted sexual and excretory activities in a patently offensive manner," at a time "when children were undoubtedly in the audience" was indecent and prohibited by 18 U.S.C. § 1464.[17] Pursuant to the Court's decision, whether any such material is "patently offensive" is determined by "contemporary community standards for the broadcast medium."[18] The Court noted that indecency is "largely a function of context — it cannot be judged in the abstract."[19]

The Commission's order in the *Pacifica* case relied partially on a spectrum scarcity argument; i.e., that there is a scarcity of spectrum space so the government must license the use of such space in the public interest, and partially on "principles analogous to those found in the law of nuisance."[20] The Commission noted that public nuisance law generally aims to channel the offensive behavior rather than to prohibit it outright. For example, in the context of broadcast material, channeling would involve airing potentially offensive material at times when children are less likely to be in the audience. In 1987, the Commission rejected the spectrum scarcity argument as a sufficient basis for its regulation of broadcast indecency, but noted that it would continue to rely upon the validity of the public nuisance rationale, including channeling of potentially objectionable material.[21] However, in its 1987 order, the Commission also stated that channeling based on a specific time of day was no longer a sufficient means to ensure that children were not in the audience when indecent material aired and warned licensees that indecent material aired after 10 p.m. would be actionable.[22] The Commission further clarified its earlier *Pacifica* order, noting that indecent language was not strictly limited to the seven words at issue in the original broadcast in question, and that repeated use of those words was not necessary to find that material in question was indecent.[23] The Commission's 1987 orders were challenged by parties alleging that the Commission had changed its indecency standard and that the new standard was unconstitutional. In *Action for Children's Television v. Federal Communications Commission (ACT I),* the United States Court of Appeals for the District of Columbia Circuit upheld the standard used by the Commission to determine whether broadcast material was indecent, but it vacated the Commission's order with respect to the channeling of indecent material for redetermination "after a full and fair hearing of the times at which indecent material may be broadcast."[24]

Following the court's decision in *Action for Children's Television (ACT I)*, a rider to the Commerce, Justice, State FY89 Appropriations Act required the FCC to promulgate regulations to ban indecent broadcasts 24 hours a day.[25] The Commission followed the congressional mandate and promulgated regulations prohibiting all broadcasts of indecent material.[26] The new regulations were challenged, and the United States Court of Appeals for the District of Columbia Circuit vacated the Commission's order.[27] In so doing, the court noted that in *ACT I* it held that Commission "must identify some reasonable period of time during which indecent material may be broadcast," thus precluding a ban on such broadcasts at all times.[28]

In 1992, Congress enacted the Public Telecommunications Act of 1992, which required the FCC to promulgate regulations to prohibit the broadcasting of indecent material from 6 a.m. to midnight, except for broadcasts by public radio and television stations that go off the air at or before midnight, in which case such stations may broadcast indecent material beginning at 10 p.m.[29] The Commission promulgated regulations as mandated in the act.[30] The new regulations were challenged, and a three-judge panel of the United States Court of Appeals for the District of Columbia Circuit subsequently vacated the Commission's order implementing the act and held the underlying statute unconstitutional.[31] In its order implementing the act, the FCC set forth three goals to justify the regulations: (1) ensuring that parents have an opportunity to supervise their children's listening and viewing of over-the-air broadcasts; (2) ensuring the well being of minors regardless of supervision; and (3) protecting the right of all members of the public to be free of indecent material in the privacy of their homes.[32] The court rejected the third justification as "insufficient to support a restriction on the broadcasting of constitutionally protected indecent material," but accepted the first two as compelling interests.[33] Despite the finding of compelling interests in the first two, the court found that both Congress and the FCC had failed "to tailor their efforts to advance these interests in a sufficiently narrow way to meet constitutional standards."[34]

Following the decision of the three-judge panel, the Commission requested a rehearing *en banc*.[35] The case was reheard on October 19, 1994, and, on June 30, 1995, the full court of appeals held the statute unconstitutional insofar as it prohibited the broadcast of indecent material between the hours of 10 p.m. and midnight on nonpublic stations.[36] In so doing, the court held that while the channeling of indecent broadcasts between midnight and 6 a.m. "would not unduly burden the First Amendment," the distinction drawn by Congress between public and non-public broadcasters "bears no apparent relationship to the compelling government interests that [the restrictions] are intended to serve."[37] The court remanded the regulations to the FCC with instructions to modify the regulations to permit the broadcast of indecent material on all stations between 10 p.m and 6 a.m.

CURRENT REGULATIONS AND ANALYSIS

Following the decision in *ACT III*, the Commission modified its indecency regulations to prohibit the broadcast any material which is indecent on any day between 6 a.m. and 10 p.m.[38] The newly modified regulations became effective August 28, 1995.[39] These regulations have been enforced primarily with respect to radio broadcasts and thus have been applied to indecent language rather than to images.[40] Broadcasts deemed indecent are subject to a forfeiture of up to $32,500 per violation.[41] Recently, the FCC started considering each utterance of an indecent word as a separate violation, rather than viewing the entire program as a single violation, which could lead to fines in excess of $32,500.[42]

To determine whether broadcast material is in fact indecent, the Commission must make two fundamental determinations: (1) that the material alleged to be indecent falls within the subject matter scope of the definition of indecency — the material in question must describe or depict sexual or excretory organs or activities; and (2) that the broadcast is patently offensive as measured by contemporary community standards for the broadcast medium.[43] If the material in question does not fall within the subject matter scope of the indecency definition,[44] or if the broadcast occurred during the "safe harbor" hours (between 10 p.m. and 6 a.m.), the complaint is usually dismissed.[45] However, if the Commission determines that the complaint meets the subject matter requirements and was aired outside the "safe harbor" hours, the broadcast in question is evaluated for patent offensiveness.[46] The Commission notes that in determining whether material is patently offensive, the full context is very important, and that such determinations are highly fact-specific.

The Commission has identified three factors that have been significant in recent decisions in determining whether broadcast material is patently offensive:

> (1) the explicitness or graphic nature of the description or depiction of sexual or excretory organs or activities; (2) whether the material dwells on or repeats at length descriptions of sexual or excretory organs or activities; (3) whether the material appears to pander or is used to titillate, or whether the material appears to have been presented for its shock value.[47]

An overview and analysis of cases addressing each of these factors follows.

Explicitness or Graphic Nature of Material

Generally, the more explicit or graphic the description or depiction, the greater the likelihood that the material will be deemed patently offensive and therefore indecent. For example, the Commission imposed a forfeiture on a university radio station for airing a rap song that included a line depicting anal intercourse.[48] In that case, the Commission determined that the song described sexual activities in graphic terms that were patently offensive and therefore indecent. Since the song was broadcast in the mid- afternoon, there was a reasonable risk that children were in the audience, thus giving rise to the Commission's action.[49]

Broadcasts need not be as graphic as the song in the above case to give rise to the imposition of an FCC forfeiture. Broadcasts consisting of double entendres or innuendos may also be deemed indecent if the "sexual or excretory import is unmistakable."[50] The FCC issued a notice of apparent liability and imposed a forfeiture on several stations for airing a song that included the following lines: "I whipped out my Whopper and whispered, Hey, Sweettart, how'd you like to Crunch on my Big Hunk for a Million Dollar Bar? Well, she immediately went down on my Tootsie Roll and you know, it was like pure Almond Joy."[51] The Commission determined that the material was indecent even though it used candy bar names to substitute for sexual activities. In one notice concerning the broadcast of the song, the Commission stated that "[w]hile the passages arguably consist of double entendre and indirect references, the language used in each passage was understandable and clearly capable of specific sexual meaning and, because of the context, the sexual import was inescapable."[52] The nature of the lyrics, coupled with the fact that the song aired between 6 a.m. and 10 a.m., gave rise to the imposition of a forfeiture.

Dwelling or Repetition of Potentially Offensive Material

Repetition of and persistent focus on a sexual or excretory activity could "exacerbate the potential offensiveness of broadcasts."[53] For example, the FCC issued a notice of apparent liability and imposed a forfeiture on a radio station that broadcast an extensive discussion of flatulence and defecation by radio personality "Bubba, the Love Sponge."[54] Though the broadcast did not contain any expletives, the Commission found that the material dwelt on excretory activities and therefore was patently offensive.

While repetition can increase the likelihood that references to sexual or excretory activities are deemed indecent, where such references have been made in passing or are fleeting in nature, the Commission has found that the reference was not indecent even when profanity has been used.[55] For example, the Commission determined that the following phrase — "The hell I did, I drove mother-f[***]er, oh." — uttered by an announcer during a radio morning show, was not indecent.[56] The Commission declined to take action regarding the broadcast because it contained only a "fleeting and isolated utterance ... within the context of live and spontaneous programming."[57] Certain fleeting references may, however, be found indecent where other factors contribute to the broadcast's patent offensiveness. For example, the Commission has imposed forfeitures on stations for airing jokes that refer to sexual activities with children.[58]

Pandering or Titillating Nature of Material

In determining whether broadcast material is indecent, the Commission also looks to the purpose for which the material is being presented. Indecency findings generally involve material that is presented in a pandering or titillating nature, or material that is presented for the shock value of its language. For example, the Commission deemed a radio call-in survey about oral sex to be indecent based in part on the fact that the material was presented in a pandering and titillating manner.[59]

Whether a broadcast is presented in a pandering or titillating manner depends on the context in which the potentially indecent material is presented. Explicit images or graphic language does not necessarily mean that the broadcast is being presented in a pandering or titillating manner. For example, the Commission declined to impose a forfeiture on a television station for airing portions of a high school sex education class that included the use of "sex organ models to demonstrate the use of various birth control devices."[60] In dismissing the complaint, the Commission held that "[a]lthough the program dealt with sexual issues, the material presented was clinical or instructional in nature and not presented in a pandering, titillating, or vulgar manner."[61]

GOLDEN GLOBE AWARDS DECISION

As noted above, on March 18, 2004, the Federal Communications Commission overturned an earlier decision by the Commission's Enforcement Bureau regarding the broadcast of the word "f[***]ing" during the 2003 Golden Globe Awards. In the earlier decision, the Enforcement Bureau had found that the broadcast of the program including the utterance did not violate federal restrictions regarding the broadcast of obscene and indecent material.[62] The Bureau dismissed the complaints primarily because the language in question did not describe or depict sexual or excretory activities or organs.

In its March 18 *Memorandum Opinion and Order*, the full Commission concluded that the broadcast of the Golden Globe Awards did include material that violated prohibitions on the broadcast of indecent and profane material.[63] In reversing the Bureau, the Commission determined that the "phrase at issue is within the scope of our indecency definition because it does depict or describe sexual activities."[64] Although the Commission "recognize[d] NBC's argument that the 'FWord' here was used 'as an intensifier,'" it nevertheless concluded that, "given the core meaning of the 'F-Word,' any use of that word or a variation, in any context, inherently has a sexual connotation, and therefore falls within the first prong of our indecency definition."[65]

Upon finding that the phrase in question fell within the first prong of the definition of "indecency," the Commission turned to the question of whether the broadcast was patently offensive under contemporary community standards for the broadcast medium. The Commission determined that the broadcast was patently offensive, noting that "[t]he 'F-Word' is one of the most vulgar, graphic and explicit descriptions of sexual activity in the English language," and that "[t]he use of the 'FWord' here, on a nationally telecast awards ceremony, was shocking and gratuitous."[66] The Commission also rejected "prior Commission and staff action [that] have indicated that isolated or fleeting broadcasts of the 'F-Word' such as that here are not indecent or would not be acted upon," concluding "that any such interpretation is no longer good law."[67] The Commission further clarified its position, stating "that the mere fact that specific words or phrases are not sustained or repeated does not mandate a finding that material that is otherwise patently offensive to the broadcast medium is not indecent."[68]

In addition to the determination that the utterance of the word "f[***]ing" during the Golden Globe Awards was indecent, the Commission also found, as an independent ground for its decision, that use of the word was "profane" in violation of 18 U.S.C. 1464.[69] In making this determination, the Commission cited dictionary definitions of "profanity" as "'vulgar, irreverent, or coarse language,'"[70] and a Seventh Circuit opinion stating that "profanity" is "'construable as denoting certain of those personally reviling epithets naturally tending to provoke violent resentment or denoting language so grossly offensive to members of the public who actually hear it as to amount to a nuisance.'"[71] The Commission acknowledged that its limited case law regarding profane speech has focused on profanity in the context of blasphemy, but stated that it would no longer limit its definition of profane speech in such manner. Pursuant to its adoption of this new definition of "profane," the Commission stated that, depending on the context, the "'F-Word' and those words (or variants thereof) that are as highly offensive as the 'F-Word'" would be considered "profane" if broadcast between 6 a.m. and 10 p.m.[72] The Commission noted that other words would be considered on a case-by-case basis.

SUPER BOWL HALFTIME SHOW DECISION

As noted above, on September 22, 2004, the FCC released a *Notice of Apparent Liability for Forfeiture* imposing a $550,000 forfeiture on several Viacom-owned CBS affiliates for the broadcast of the Super Bowl XXXVIII halftime show on February 1, 2004, in which a performer's breast was exposed.[73] The Commission determined that the show, which was aired at approximately 8:30 p.m. Eastern Standard Time, violated its restrictions on the broadcast of indecent material.

In its analysis, the Commission determined that since the broadcast included a performance that culminated in "on-camera partial nudity," and thus satisfied the first part of the indecency analysis, further scrutiny was warranted to determine whether the broadcast was "patently offensive as measured by contemporary community standards for the broadcast medium."[74] The Commission found that the performance in question was "both explicit and graphic," and rejected the licensees' contention that since the exposure was fleeting, lasting only 19/32 of a second, it should not be deemed indecent.[75] In determining whether the material in question was intended to "pander to, titillate and shock the viewing audience," the Commission noted that the performer's breast was exposed after another performer sang, "gonna have you naked by the end of this song."[76] The Commission found that the song lyrics, coupled with simulated sexual activities during the performance and the exposure of the breast, indicated that the purpose of the performance was to pander to, titillate and shock the audience, and the fact that the actual exposure of the breast was brief, as noted above, was not dispositive.[77]

The Commission ordered each Viacom-owned CBS affiliate to pay the statutory maximum forfeiture of $27,500 for the broadcast, for a total forfeiture of $550,000. The forfeiture was imposed on the Viacom-owned affiliates because of Viacom's participation in and planning of the Super Bowl halftime show with MTV networks, another Viacom subsidiary.[78]

Following the issuance of the *Notice of Apparent Liability for Forfeiture*, the affiliates are "afforded a reasonable period of time (usually 30 days from the date of the notice) to show, in writing, why a forfeiture penalty should not be imposed or should be reduced, or to pay the forfeiture."[79] CBS filed an opposition to the Notice of Apparent Liability on November 5, 2004. The opposition challenged the forfeiture on various grounds, including that the test for indecency was not met and that the forfeiture violates the First Amendment. The Commission may now issue an order cancelling or reducing the proposed forfeiture, or it may require that the original amount be paid in full by a certain date.[80] This order, constituting a final agency action, may subsequently be appealed in federal court.[81]

OTHER RECENT ENFORCEMENT ACTIONS

In addition to the Commission's forfeiture proceedings discussed above, the Commission has recently imposed forfeitures on a number of radio stations for broadcast indecency.[82] We now discuss two of its recent high-profile actions.

Infinity Broadcasting

On October 2, 2003, the Commission issued a *Notice of Apparent Liability* to Infinity Broadcasting for airing portions of the "Opie & Anthony Show" during which the hosts conducted a contest entitled "Sex for Sam" which involved couples having sex in certain "risky" locations throughout New York City in an effort to win a trip.[83] The couples, accompanied by a station employee, were to have sex in as many of the designated locations as possible. They were assigned points based on the nature of the location and the activities in which they engaged. The station aired discussions between the hosts of the show and the station employee accompanying the couples which consisted of descriptions of the sexual activities of the participating couples and the locations in which they engaged in sexual activities. One discussion involved an description of a couple apparently engaging in sexual activities in St. Patrick's Cathedral.

The Commission determined that the broadcast made "graphic and explicit references to sexual and excretory organs and activity" despite the fact that colloquial terms, rather than explicit or graphic terms, were used in the descriptions. The Commission found that "[t]o the extent that the colloquial terms that the participants used to describe organs and activities could be described as innuendo rather than as direct references, they are nonetheless sufficient to render the material actionably indecent because the 'sexual [and] excretory import' of those references was 'unmistakable.'"[84] The Commission also found that the hosts of the show "dwelled at length on and referred repeatedly to sexual or excretory activities and organs," and that "the descriptions of sexual and excretory activity and organs were not in any way isolated and fleeting."[85]

On November 23, 2004, the FCC entered into a consent decree with Infinity regarding the Opie & Anthony *NAL*.[86] Pursuant to the decree, Infinity, a subsidiary of Viacom, agreed to make a voluntary contribution to the United States Treasury in the amount of $3.5 million and to adopt a company-wide compliance plan for the purpose of preventing the broadcast of indecent material. As part of the companywide plan, Viacom agreed to install delay systems to edit "potentially problematic" live programming and to conduct training with respect to indecency regulations for all of its on-air talent and employees who participate in programming decisions.

Clear Channel Broadcasting

On January 27, 2004, the Commission issued a *Notice of Apparent Liability* to Clear Channel Broadcasting for repeated airings of the "Bubba, the Love Sponge" program which included indecent material.[87] The Commission found that all the broadcasts in question involved "conversations about such things as oral sex, penises, testicles, masturbation, intercourse, orgasms and breasts."[88] The Commission determined that each of the broadcasts in question contained "sufficiently graphic and explicit references," which were generally repeated throughout the broadcast in a pandering and titillating manner.

In one broadcast, the station aired a segment involving skits in which the voices of purported cartoon characters talk about drugs and sex.[89] The skits were inserted between advertisements for Cartoon Network's Friday-night cartoons. The Commission determined that "the use of cartoon characters in such a sexually explicit manner during hours of the day when children are likely to be listening is shocking and makes this segment patently offensive."[90] The Commission also cited the "calculated and callous nature of the stations' decision to impose this predictably offensive material upon young, vulnerable listeners" as "weighing heavily" in its determination.[91]

On April 8, 2004, the Commission released another *Notice of Apparent Liability* against Clear Channel Communications for airing allegedly indecent material during the "Howard Stern Show."[92] For the first time, the Commission sought to impose separate statutory maximum forfeitures for each indecent utterance during the program in question, rather than imposing a single fine for the entire program.[93]

The Commission entered into a consent decree with Clear Channel on June 9, 2004. The decree requires Clear Channel to make a "voluntary contribution" of $1.75 million to the United States Treasury and outlines "a company-wide compliance plan for the purpose of preventing the broadcast over radio or television of material violative of the indecency laws."[94] As part of the compliance plan, Clear Channel will "conduct training on obscenity and indecency for all on-air talent and employees who materially participate in programming decisions, which will include tutorials regarding material that the FCC does not permit broadcasters to air."[95] The plan also requires Clear Chanel to suspend any employee accused of airing, or who materially participates in the decision to air, obscene or indecent material while an investigation is conducted following the issuance of a *Notice of Apparent Liability.* Such employees will be terminated without delay if the *NAL* results in enforcement action by the FCC.

CONGRESSIONAL RESPONSE

108th Congress

Several bills addressing broadcast indecency were introduced during the 108th Congress, but none were enacted.

In response to the FCC's initial decision regarding the Golden Globe Awards broadcast, Representative Ose introduced legislation that would have amended section 1464 of title 18 to define "profane," as used in that section, to include any use of eight specific words or phrases, in any grammatical form.[96]

Legislation was also introduced, by Representative Upton, that would have increased the FCC's penalties for broadcasting obscene, indecent, and profane language. H.R. 3717, 108th Congress, would have increased the penalties to $275,000 for each violation or each day of a continuing violation, with the total amount assessed for any continuing violation not to exceed $3 million for any single act or failure to act. The bill specifically mentioned obscene, indecent, and profane language, but did not appear to impose the increased penalties for broadcast images that were deemed indecent.

On March 3, 2004, the House Committee on Energy and Commerce held a fullcommittee markup of H.R. 3717. The Committee approved an amendment in the nature of a substitute and ordered the bill to be reported to the full House.[97] The amendment would have increased the maximum penalty for the broadcast of indecent material to $500,000 for each violation with no aggregate maximum. Under the amendment, the increased forfeiture penalties could have also been imposed on nonlicensees, such as individuals who utter obscene, indecent, or profane material. In determining the appropriate amount of the forfeiture, the bill directed the Commission to consider a number of factors, including whether the material in question was live or recorded, scripted or unscripted, and the size of the viewing audience. The Commission would have been required to act on allegations of broadcast indecency within 180 days after the receipt of the allegation.

In addition to the increased penalties, the legislation also would have provided additional nonmonetary penalties for the broadcast of indecent material. If the Commission determined that a licensee had broadcast obscene, indecent, or profane material, the Commission would have been able to require the licensee to broadcast public service announcements that serve the educational and informational needs of children, and such announcements could have been required to reach an audience that was up to five times the size of the audience that was estimated to have been reached by the obscene, indecent, or profane material. The legislation also directed the Commission to take into consideration whether the broadcast of obscene, indecent, or profane material demonstrated a lack of character or other qualifications to operate a station. The broadcast of such material would have also been considered a serious violation for purposes of license renewal determinations. Finally, if during the term of the license, a broadcast licensee was the subject of three or more proceedings regarding violations of indecency prohibitions, the Commission would have been required to commence a proceeding to consider whether the license should be revoked.

The House passed H.R. 3717 on March 11, 2004, with two additional amendments.[98] The first amendment would have preserved a licensee's right to appeal a forfeiture before it could have been considered during a license application or renewal proceeding, or used in a license revocation proceeding initiated after the licensee's third indecency violation. The second amendment approved by the House would have required the GAO to study the number of indecency complaints received by the FCC and the number of those complaints that result in final agency action by the Commission; the amount of time taken by the Commission to respond to a complaint; the mechanisms established by the Commission to receive, investigate, and respond to complaints; and whether complainants to the Commission are adequately informed by the Commission of the responses to their complaints. The amendment gave GAO one year to complete the study and report to Congress on its findings.

On February 9, 2004, Senator Brownback introduced S. 2056, which was virtually identical to H.R. 3717, as introduced. On March 9, 2004, the Senate Committee on Commerce, Science and Transportation held a full committee markup of the bill, and ordered the bill to be reported favorably with amendments.[99] As reported, the bill included a number of provisions similar to those in H.R. 3717. Both bills would have increased forfeiture amounts for broadcast indecency, but unlike the House bill, which included an initial increase to $500,000 per violation, the Senate bill would have provided for an initial increase to $275,000 per violation, with additional increases, up to $500,000, for subsequent violations. Penalties could have been doubled if the Commission determined that certain aggravating factors were met. Like the House bill, the Senate bill included a provision that would have required the FCC to commence a license revocation hearing if a broadcast licensee had been the subject of three proceedings regarding indecency violations. The Senate bill would have also required the Commission to act on any allegations of broadcast indecency within 270 days.

Additional amendments to S. 2056 would have directed the Commission to consider a broadcaster's ability to pay forfeitures based on factors such as revenues and market size when determining the amount of the forfeiture, and would have invalidated the FCC's media ownership rules pending a GAO review of the relationship between horizontal and vertical consolidation of media companies and violations of indecency prohibitions.[100]

Another Senate bill aimed at increasing penalties for broadcast indecency was introduced by Senator Miller on March 1, 2004. S. 2147 would have allowed the Commission to impose a forfeiture of "25 cents times the number of individuals who witnessed or heard the broadcast as determined by a viewership rating service selected by the Commission" for the broadcast of obscene, indecent, or profane language. The forfeiture could have been imposed on broadcast licensees, any applicant for a broadcast license, or any other company or individual that had participated in the broadcast, including producers, general managers, performers, and networks.

On June 22, 2004, the Senate approved an amendment to the National Defense Authorization Act (S. 2400) which included provisions similar to those in S. 2056, as reported by the Senate Commerce, Science, and Transportation Committee on March 9, 2004. The amendment would have increased the maximum forfeiture imposed by the FCC to $275,000 per violation, with a cap of $3 million for a continuing violation.[101] The Senate's amendment to the National Defense Authorization Act was not adopted by the conference committee and was not included in the conference report as approved by both the House and Senate on October 9, 2004.[102]

In addition, S. 1264, 108th Congress, the FCC Reauthorization Act of 2003, as reported by the Committee on Commerce on September 3, 2003 (prior to the FCC Enforcement Bureau's decision regarding the Golden Globe Awards broadcast), would have, in section 11, required that "the broadcast of obscene or indecent matter from more than 1 individual during the same program" be considered separate violations, and that the FCC, unless it determined it not to be in the public interest, revoke the station license or construction permit of any broadcast station licensee or permittee that violates 18 U.S.C. § 1464.

None of the above indecency provisions was enacted into law.

Finally, several resolutions were introduced to express disapproval of the FCC Enforcement Bureau's decision regarding the Golden Globe Awards broadcast: H.Res. 482,

H.Res. 500, and S.Res. 283, which the Senate passed on December 9, 2003. None of the House resolutions were acted upon.

109th Congress

Legislation addressing broadcast indecency has been introduced during the 109th Congress. In general, these bills are similar to legislation introduced during the 108th Congress.

H.R. 310, 109th Congress, entitled the Broadcast Decency Enforcement Act of 2005, would increase penalties for the broadcast of obscene, indecent, or profane material to a maximum of $500,000 for each violation, and would provide penalties for nonlicensees, such as artists and performers.[103] In addition to the increased forfeiture amount, the Federal Communications Commission would be able to require licensees to broadcast public service announcements that serve the educational and informational needs of children and require such announcements to reach an audience up to five times greater than the size of the audience that is estimated to have been reached by the obscene, indecent or profane broadcast. Also, any violation of the Commission's indecency regulations could be considered when determining whether to grant or renew a broadcast license, and three or more indecency violations would trigger a license revocation proceeding. The legislation would also require the Commission to act upon allegations of indecency within 180 days of the receipt of the allegation.

On February 9, 2005, the House Committee on Energy and Commerce reported H.R. 310 favorably without amendment. The House passed H.R. 310 on February 16, with one amendment. The amendment made several nonsubstantive technical changes and added a section requiring the FCC to revise its policy statement regarding broadcast indecency within nine months of the date of enactment of the legislation and at least once every three years thereafter.[104]

H.R. 1420, 109th Congress, entitled the Families for ED Advertising Decency Act, would require the Federal Communications Commission to revise its interpretations of, and enforcement policies concerning the broadcast of indecent material to treat as indecent any advertisement for a medication for the treatment of erectile dysfunction, thus prohibiting the airing of such advertisements between the hours of 6 a.m. and 10 p.m.

H.R. 1440, 109th Congress, entitled the Stamp Out Censorship Act of 2005, would prohibit the Federal Communications Commission from imposing forfeitures for indecency on nonbroadcast programming.

S. 193, 109th Congress, also entitled the Broadcast Decency Enforcement Act of 2005, would increase the maximum forfeiture for the broadcast of obscene, indecent, or profane material to $325,000 for each violation, with a cap of $3,000,000 for any single act or failure to act.

S. 616, 109th Congress, entitled the Indecent and Gratuitous and Excessively Violent Programming Control Act of 2005, would, *inter alia,* require the Federal Communications Commission to study the effectiveness of measures used by multichannel video programming distributors to protect children from exposure to indecent and violent programming. If the Commission were to determine that such measures were not effective, the legislation directs the Commission to initiate a rulemaking proceeding to adopt measures to protect children from indecent programming carried by multichannel video programming distributors during the hours when children are reasonably likely to be a substantial portion of the television audience. The bill would also increase the penalty for the broadcast of obscene, indecent, or profane language or images to $500,000, with a maximum forfeiture for any violations occurring in a given 24-hour period set at $3,000,000. The Commission would be required to take certain factors into consideration when imposing a forfeiture, and would be required to conduct public hearings or forums prior to the imposition of a forfeiture. Additionally, broadcast licensees would be allowed to preempt programming from a network organization that it deems obscene, and all television and radio broadcast licensees and multichannel video programming distributors would be required to provide a warning of the specific content of each recorded or scripted program it broadcasts.

Would Prohibiting the Broadcast of "Indecent" Words Regardless of Context Violate the First Amendment?

In 1978, in *Federal Communications Commission v. Pacifica Foundation*, the Supreme Court upheld, against a First Amendment challenge, an action that FCC took against a radio station for broadcasting a recording of George Carlin's "Filthy Words" monologue at 2 p.m.[105] The Court has not decided a case on the issue of "indecent" speech on broadcast radio or television since then, but it did cite *Pacifica* with approval in 1997, when, in *Reno v. ACLU*, it contrasted regulation of the broadcast media with regulation of the Internet.[106] Nevertheless, the Court in *Reno* did not *hold* that *Pacifica* remains good law, and arguments have been made that the proliferation of cable television channels has rendered archaic *Pacifica*'s denial of full First Amendment rights to broadcast media.

Even if *Pacifica* remains valid in this respect, *Pacifica* did not hold that the First Amendment permits the ban either of an occasional expletive on broadcast media, or of programs that would not be likely to attract youthful audiences, even if such programs contain "indecent" language. On these points, Justice Stevens wrote for the Court in *Pacifica*:

> It is appropriate, in conclusion, to emphasize the narrowness of our holding. This case does not involve a two-way radio conversation between a cab driver and a dispatcher, or a telecast of an Elizabethan comedy. We have not decided that an occasional expletive in either setting would justify any sanction. ... The time of day was emphasized by the Commission. The content of the program in which the language is used will also affect the composition of the audience. ... [107]

In a footnote to the last sentence of this quotation, the Court added: "Even a prime-time recitation of Geoffrey Chaucer's Miller's Tale would not be likely to command the attention of many children. ... "[108] At the same time, Justice Stevens acknowledged that the Carlin monologue has political content: "The monologue does present a point of view; it attempts to show that the words it uses are 'harmless' and that our attitudes toward them are 'essentially silly.' The Commission objects, [however,] not to this point of view, but to the way in which it is expressed."[109] The Court commented: "If there were any reason to believe that the Commission's characterization of the Carlin monologue as offensive could be traced to its political content — or even to the fact that it satirized contemporary attitudes about four-letter words — First Amendment protection might be required."[110]

There appears to be some tension between this comment and the Court's remark about Chaucer, as any attempt to censor Chaucer would presumably also be based not on its ideas but on the way its ideas are expressed. But, as noted above, the Court's remark about Chaucer was a footnote to its comment that "[t]he content of the program in which the language is used will also affect the composition of the audience. ... " Therefore, the difference that Justice Stevens apparently perceived between Chaucer and Carlin was that, even if both have literary, artistic, or political value, only the latter would be likely to attract a youthful audience. Arguably, then, *Pacifica* would permit the censorship, during certain hours, of the broadcast even of works of art that are likely to attract a youthful audience.[111]

If so, this would be contrary to the Court's opposition, in other contexts, to the censorship of works of art. The Court has held that even "materials [that] depict or describe patently offensive 'hard core' sexual conduct," which would otherwise be obscene, may not be prohibited if they have "serious literary, artistic, political, or scientific value."[112] In addition, the "harmful to minors" statutes of the sort that the Supreme Court upheld in *Ginsberg v. New York* generally define "harmful to minors" to parallel the Supreme Court's definition of "obscenity," and thus prohibit distributing to minors only material that lacks serious value for minors.[113] This suggests that, if the FCC or Congress prohibited the broadcast during certain hours of "indecent" words regardless of context, the Court might be troubled by the prohibition's application to works with serious value, even though *Pacifica* allowed the censorship of Carlin's monologue, despite its apparently having serious value.

Yet, as noted, Justice Stevens' expressed a distinction in *Pacifica* between a point of view and the way in which it is expressed, and, though a majority of the justices did not join the part of the opinion that drew this distinction, a majority of the justices, by concurring in *Pacifica*'s holding, indicated that the political (or literary or artistic) content of Carlin's monologue did not prevent its censorship during certain hours on broadcast radio and television. Therefore, it appears that, in deciding the constitutionality of an FCC or a congressional action prohibiting the broadcasting, during certain hours, of material with "indecent" words, the Court might be troubled by its application to works with serious value only if those works would, like Chaucer's, not likely attract a substantial youthful audience.

In sum, the Court did not hold that the FCC could prohibit an occasional expletive, and did not hold that the FCC could prohibit offensive words in programs — even prime-time programs — that children would be unlikely to watch or listen to. The Court did not hold that the FCC could *not* take these actions, as the question whether it could was not before the Court. But the Court's language quoted above renders *Pacifica* of uncertain precedential value in deciding whether a ban, during certain hours, on the broadcast of "indecent" words regardless of context would be constitutional.

In the "Filthy Words" monologue, as the Supreme Court described it, George Carlin "began by referring to his thoughts about 'the words you couldn't say on the public, ah, airwaves, um, the ones you definitely wouldn't say, ever.' He proceeded to list those words and repeat them over and over in a variety of colloquialisms." The FCC, at the time, used essentially the same standard for "indecent" that it uses today: "[T]he concept of 'indecent' is intimately connected with the exposure of children to language that describes, in terms patently offensive as measured by contemporary community standards for the broadcast medium, sexual or excretory activities and organs. ... "[114]

Most of Carlin's uses of the "filthy words," it appears from reading his monologue, which is included as an appendix to the Court's opinion, seem designed to show the words' multiple uses, apart from describing sexual or excretory activities or organs. Nevertheless, "the Commission concluded that certain words depicted sexual or excretory activities in a patently offensive manner. ... "[115] Therefore, one might argue that, even if, under *Pacifica*, the First Amendment does not protect, during certain hours, the use on broadcast media of words that depict sexual or excretory activities in a patently offensive manner, it nevertheless might protect the use of those same words "as an adjective or expletive to emphasize an exclamation" (to quote the FCC Enforcement Bureau's opinion in the Bono case).

A counterargument might be that, in *Pacifica*, the Court noted that "the normal definition of 'indecent' merely refers to nonconformance with accepted standards of morality."[116] This suggests the possibility that the Court would have ruled the same way in *Pacifica* if the FCC had defined "indecent" loosely enough to include the use of a patently offensive word "as an adjective or expletive to emphasize an exclamation." But this is speculative, as the Court did not so rule. Further, as noted above, Court emphasized the narrowness of its holding, noting that it had "not decided that an occasional expletive ... would justify any sanction. ... "

On what basis did the Court in *Pacifica* find that the FCC's action did not violate the First Amendment? In Part IV-C of opinion, which was joined by a majority of the justices, Justice Stevens wrote:

> [O]f all forms of communication, it is broadcasting that has received the most limited First Amendment protection. Thus, although other speakers cannot be licensed except under laws that carefully define and narrow official discretion, a broadcaster may be deprived of his license and his forum if the Commission decides that such an action would serve "the public interest, convenience, and necessity." Similarly, although the First Amendment protects newspaper publishers from being required to print the replies of those whom they criticize, Miami Herald Publishing Co. v. Tornillo, 418 U.S. 241, it affords no such protection to broadcasters; on the contrary, they must give free time to the victims of their criticism. Red Lion Broadcasting Co. v. FCC, 395 U.S. 367.
>
> The reasons for these distinctions are complex, but two have relevance to the present case. First, the broadcast media have established a uniquely pervasive presence in the lives of all Americans. Patently offensive, indecent material presented over the airwaves confronts the citizen, not only in public, but in the privacy of the home, where the individual's right to be left alone plainly outweighs the First Amendment rights of an intruder. Rowan v. Post Office Dept., 397 U.S. 728. ... To say that one may avoid further offense by turning off the radio when he hears indecent language is like saying that the remedy for an assault is to run away after the first blow.
>
> Second, broadcasting is uniquely accessible to children, even those too young to read. ... Bookstores and motion picture theaters ... may be prohibited from making indecent material available to children. We held in Ginsberg v. New York, 390 U.S. 629, that the government's interest in the "well-being of its youth" and in supporting "parents' claim to authority in their own household" justified the regulation of otherwise protected expression. ... 117

In sum, the Court held that, on broadcast radio and television, during certain times of day, certain material may be prohibited because (1) it is patently offensive and indecent, and (2) it threatens the well-being of minors and their parents' authority in their own household. This raises the question of the extent to which the Court continues to allow the government (1) to treat broadcast media differently from other media, and (2) to censor speech on the ground that it is patently offensive and indecent, or threatens the well-being of minors and their parents' authority in their own household.

Broadcast Media

In *Red Lion Broadcasting Co. v. FCC*, which the Court cited in the above quotation from *Pacifica*, the Court upheld the FCC's "fairness doctrine," which "imposed on radio and television broadcasters the requirement that discussion of public issues be presented on broadcast stations, and that each side of those issues must be given fair coverage."[118] The reason that the Court upheld the imposition of the fairness doctrine on broadcast media, though it would not uphold its imposition on print media, is that "[w]here there are substantially more individuals who want to broadcast than there are frequencies to allocate, it is idle to posit an unabridgeable First Amendment right to broadcast comparable to the right of every individual to speak, write, or publish."[119] "Licenses to broadcast," the Court added, "do not confer ownership of designated frequencies, but only the temporary privilege of using them. 47 U.S.C. § 301. Unless renewed, they expire within three years. 47 U.S.C. § 307(d). The statute mandates the issuance of licenses if the 'public convenience, interest, or necessity will be served thereby.' 47 U.S.C. § 307(a)."[120]

The Court in *Red Lion* then noted:

> It is argued that even if at one time the lack of available frequencies for all who wished to use them justified the Government's choice of those who would best serve the public interest ... this condition no longer prevails so that continuing control is not justified. To this there are several answers. Scarcity is not entirely a thing of the past.[121]

With the plethora of cable channels today, has spectrum scarcity now become a thing of the past? In Turner Broadcasting System, Inc. v. FCC, the Court held that the scarcity rationale does not apply to cable television:

> [C]able television does not suffer from the inherent limitations that characterize the broadcast medium ... [S]oon there may be no practical limitation on the number of speakers who may use the cable medium. Nor is there any danger of physical interference between two cable speakers attempting to use the same channel. In light of these fundamental technological differences between broadcast and cable transmission, application of a more relaxed standard of scrutiny adopted in *Red Lion* and the other broadcast cases is inapt when determining the First Amendment validity of cable regulation."[122]

One might argue that, if the scarcity rationale does not apply to cable television, then it should not apply to broadcast television either, because a person who because of scarcity cannot start a broadcast channel can start a cable channel.[123] But the Court has not ruled on the question; in *Turner* it wrote: "Although courts and commentators have criticized the scarcity rationale since its inception, we have declined to question its continuing validity as support for our broadcast jurisprudence, and see no reason to do so here."[124]

In 1987, however, the FCC abolished the fairness doctrine, on First Amendment grounds, noting that technological developments and advancements in the telecommunications marketplace have provided a basis for the Supreme Court to reconsider its holding in *Red*

Lion. The FCC's decision was upheld by the U.S. Court of Appeals for the District of Columbia, and the Supreme Court declined to review the case.[125] The court of appeals did not rule on constitutional grounds, but rather concluded "that the FCC's decision that the fairness doctrine no longer served the public interest was neither arbitrary, capricious nor an abuse of discretion, and [we] are convinced that it would have acted on that finding to terminate the doctrine even in the absence of its belief that the doctrine was no longer constitutional."[126]

But, whether or not spectrum scarcity has become a thing of the past, it would apparently would not today justify governmental restrictions on "indecent" speech. This is because, subsequent to the Court in *Turner* declining to question the applicability of the scarcity rationale to broadcast media, a plurality of justices noted, in *Denver Area Educational Telecommunications Consortium, Inc. v. FCC*, that, though spectrum scarcity continued to justify the "structural regulations at issue there [in *Turner*] (the 'must carry' rules), it has little to do with a case that involves the effects of television viewing on children. Those effects are the result of how parents and children view television programming, and how pervasive and intrusive that programming is. In that respect, cable and broadcast television differ little, if at all."[127] The plurality therefore upheld a federal statute that permits cable operators to prohibit indecent material on leased access channels. Thus, it appears that the Court today would not cite spectrum scarcity to justify restrictions on "indecent" material on broadcast media, but it might cite broadcast media's pervasiveness and intrusiveness.

Subsequent to *Denver Area*, in *United States v. Playboy Entertainment Group, Inc.*, the Court held that cable television has full First Amendment protection; i.e., content-based restrictions on cable television receive strict scrutiny.[128] Thus, if, as the Court said in *Denver Area*, cable and broadcast media differ little, if at all, with respect to the regulation of "indecent" material, and, if, as the Court said in *Playboy*, cable television receives strict scrutiny, then, arguably, broadcast media would also receive strict scrutiny with regard to restrictions on "indecent" material.[129] It is possible, however, that, if cable and broadcast media differ little, then the Court might apply *Pacifica* to both broadcast and cable, rather than to neither.[130] In any event, as noted above, even if the Court were to continue to apply *Pacifica* to restrictions on broadcast media, this does not necessarily mean that it would uphold a ban on the broadcast of "indecent" language regardless of context, as *Pacifica* did not hold that an occasional expletive would justify a sanction.

Strict Scrutiny

We now consider the analysis that the Court might apply if it chooses not to apply *Pacifica* in deciding the constitutionality of a ban on the broadcast of "indecent" language regardless of context. The Court in *Pacifica*, as noted, offered two reasons why the FCC could prohibit offensive speech on broadcast media: "First, the broadcast media have established a uniquely pervasive presence in the lives of all Americans. Patently offensive, indecent material presented over the airwaves confronts the citizen, not only in public, but in the privacy of the home. ... Second, broadcasting is uniquely accessible to children, even those too young to read," and the government has an interest in the "well-being of its youth"and "in supporting 'parents' claim to authority in their own household.'" The first of these reasons apparently refers to adults as well as to children.

Ordinarily, when the government restricts speech, including "indecent" speech, on the basis of its content, the restriction, if challenged, will be found constitutional only if it satisfies "strict scrutiny."[131] This means that the government must prove that the restriction serves "to promote a compelling interest" and is "the least restrictive means to further the articulated interest."[132] The Court in *Pacifica* did not apply this test or any weaker First Amendment test, and did not explain why it did not. Its reason presumably was that the FCC's action restricted speech only on broadcast media. If, however, the Court were not to apply *Pacifica* in determining the constitutionality of a ban, during certain hours, on the broadcast of "indecent" language regardless of context, then it would apparently apply strict scrutiny.

If the Court were to apply strict scrutiny in making this determination, it seems unlikely that it would find the first reason cited in *Pacifica* — sparing citizens, including adults, from patently offensive or indecent words — to constitute a compelling governmental interest. The Court has held that the government may not prohibit the use of offensive words unless they "fall within [a] relatively few categories of instances," such as obscenity, fighting words, or words "thrust upon unwilling or unsuspecting viewers."[133]

If the Court were to apply strict scrutiny in determining the constitutionality of a ban, during certain hours, on the broadcast of "indecent" language regardless of context, it also might not find the second reason cited in *Pacifica* — protecting minors from patently offensive and indecent words and "supporting 'parents' claim to authority in their own household'" — to constitute a compelling governmental interest. When the Court considers the constitutionality of a restriction on speech, it ordinarily — even when the speech lacks full First Amendment protection and the court applies less than strict scrutiny — requires the government to "demonstrate that the recited harms are real, not merely conjectural, and that the regulation will in fact alleviate these harms in a direct and material way."[134] With respect to restrictions designed to deny minors access to sexually explicit material, by contrast, the courts appear to assume, without requiring evidence, that such material is harmful to minors, or to consider it "obscene as to minors," even if it is not obscene as to adults, and therefore not entitled to First Amendment protection with respect to minors, whether it is harmful to them or not.[135] A word used as a mere adjective or expletive, however, arguably does not constitute sexually oriented material.[136] Therefore, if a court applied strict scrutiny to decide the constitutionality of a ban, during certain hours, on the broadcast of "indecent" words regardless of context, then, in determining the presence of a compelling interest, the court might require the government to "demonstrate that the recited harms are real, not merely conjectural, and that the regulation will in fact alleviate these harms in a direct and material way." This could raise the question, not raised in *Pacifica*, of whether hearing such words is harmful to minors. More precisely, it might raise the question of whether hearing such words on broadcast radio and television is harmful to minors, even in light of the opportunities for minors to hear such words elsewhere. If the government failed to prove that hearing certain words on broadcast radio or television is harmful to minors, then a court would not find a compelling interest in censoring those words and might strike down the law.

It might still uphold the law, however, if it found that the law served the government's interest "in supporting 'parents' claim to authority in their own household,'" and that this is a compelling interest independent from the interest in protecting the well-being of minors. In *Ginsberg v. New York*, the Court referred to the state's interest in the well-being of its youth as "independent" from its interest in supporting "parents' claim to authority in their own household to direct the rearing of their children."[137] The holding in *Ginsberg*, however, did not turn on whether these interests are independent, and one might argue that they are not, because the government's interest in supporting parents lies in assisting them in protecting their children from harmful influences. If "indecent" words are not a harmful influence, then, arguably, the government has no interest, sufficient to override the First Amendment, in supporting parents in their efforts to prevent their children's access to them. It has also been argued that "a law that effectively *bans* all indecent programming ... does not facilitate parental supervision. In my view, my right as a parent has been preempted, not facilitated, if I am told that certain programming will be banned from my ... television. Congress cannot take away my right to decide what my children watch, absent some showing that my children are in fact at risk of harm from exposure to indecent programming."[138]

If the government could persuade a court that a ban, during certain hours, on the broadcast of "indecent" words regardless of context serves a compelling interest — either in protecting the well-being of minors or in supporting parents' claim to authority — the government would then have to prove that the ban was the least restrictive means to advance that interest. This might raise questions such as whether it is necessary to prohibit particular words on weekdays during school hours, solely to protect pre-school children and children who are home sick some days. In response to this question, the government could note that the broadcast in *Pacifica* was at 2 p.m. on a Tuesday, but was nevertheless considered a "time[ ] of the day when there is a reasonable risk that children may be in the audience."[139] More significantly, however, a court might find a ban too restrictive because it would prohibit the broadcast, between certain hours, of material, including works of art and other material with serious value, that would not attract substantial numbers of youthful viewers or listeners.

In conclusion, it appears that, if a court were to apply strict scrutiny to determine the constitutionality of a ban on the broadcast of "indecent" language regardless of context, then it might require the government to "demonstrate that the recited harms are real, not merely conjectural, and that the regulation will in fact alleviate these harms in a direct and material way." This would mean that the government would have to demonstrate a compelling governmental interest, such as that hearing "indecent" words on broadcast radio and television is harmful to minors, despite the likelihood that minors hear such words elsewhere, or that banning "indecent" words is necessary to support parents' authority in their own household. If the government could not demonstrate a compelling governmental interest, then the court might find the ban unconstitutional. Even if the government could demonstrate a compelling interest, a court might find the ban unconstitutional if it applied to material with serious value, at least if such material would not attract substantial numbers of youthful viewers or listeners.

Whether a court would apply strict scrutiny would depend upon whether, in light of the proliferation of cable television, it finds *Pacifica* to remain applicable to broadcast media. If a court does find that *Pacifica* remains applicable to broadcast media, then the court would be faced with questions that *Pacifica* did not decide: whether, on broadcast radio and television during hours when children are likely to be in the audience, the government may prohibit an "indecent" word used as an occasional expletive, or in material that would not attract substantial numbers of youthful viewers or listeners.

ENDNOTES

1 The FCC's indecency regulations only apply to broadcast radio and television, and not to cable television. The distinction between broadcast and cable television arises in part from the fact that the rationale for regulation of broadcast media — the dual problems of spectrum scarcity and signal interference — do not apply in the context of cable. As a result, regulation of cable television is entitled to heightened First Amendment scrutiny. *See Turner Broadcasting v. Federal Communications Commission*, 512 U.S. 622 (1994). Cable television is also distinguished from broadcast television by the fact that cable involves a voluntary act whereby a subscriber affirmatively chooses to bring the material into his or home. *See Cruz v. Ferre*, 755 F.2d 1415 (11th Cir. 1985).

2 The final section of this report ("Would prohibiting the broadcast of 'indecent' words regardless of context violate the First Amendment?") was written by Henry Cohen; the rest of the report was written by Angie A. Welborn.

3 See *In the Matter of Complaints Against Various Broadcast Licensees Regarding Their Airing of the "Golden Globe Awards" Program*, 18 F.C.C. Rcd. 19859 (2003).

4 *Id.* at 2.

5 *Id.*

6 *Id.* at 3.

7 *Id.*

8 *FCC Chairman Seeks Reversal on Profanity*, Washington Post, January 14, 2004, at E01.

9 *In the Matter of Complaints Against Various Broadcast Licensees Regarding Their Airing of the "Golden Globe Awards" Program*, File No. EB-03-IH-0110 (March 18, 2004).

10 [http://hraunfoss.fcc.gov/edocs_public/attachmatch/DOC-243435A1.pdf].

11 *Complaints Against Various Television Licensees Concerning Their February 1, 2004, Broadcast of the Super Bowl XXXVIII Halftime Show*, File No. EB-04-IH-0011 (September 22, 2004).

12 *Id.*

13 18 U.S.C. § 1464. "Radio communication" includes broadcast television, as the term is defined as "the transmission by radio of writing, signs, signals, pictures, and sounds of all kinds." 47 U.S.C. § 153(33).

14 47 U.S.C. § 503(b).

15 438 U.S. 726 (1978).

16 The United State Court of Appeals for the District of Columbia Circuit had reversed the Commission's order. *See* 556 F.2d 9 (D.C. Cir. 1977). The Commission appealed that decision to the Supreme Court. The Court granted certiorari and reversed the lower court's opinion.

17 438 U.S. at 732.

18 *Id.*

19 *Id.* at 742.

20 *Id.* at 731; *see*, *In the Matter of a Citizen's Complaint Against Pacifica Foundation Station WBAI (FM), New York, New York*, 56 F.C.C.2d 94 (1975).

21 *In the Matter of Pacifica Foundation, Inc. d/b/a Pacifica Radio Los Angeles, California*, 2 F.C.C. Rcd. 2698 (1987). Two other orders handed down the same day articulate the

Commission's clarified indecency standard. *See also In the Matter of the Regents of the University of California*, 2 F.C.C. Rcd. 2703 (1987); *In the Matter of Infinity Broadcasting Corporation of Pennsylvania*, 2 F.C.C. Rcd. 2705 (1987).

22 The Commission noted Arbitron ratings indicating that a number of children remain in the local audience well after 10 p.m. *See* 2 F.C.C. Rcd. 1698, ¶ 16.

23 2 F.C.C. Rcd. 2698, ¶¶ 12 and 15.

24 852 F.2d 1332, 1344 (1988).

25 Pub. L. 100-459, § 608.

26 *Enforcement of Prohibitions Against Broadcast Obscenity and Indecency*, 4 F.C.C. Rcd. 457 (1988).

27 *Action for Children's Television v. Federal Communications Commission (ACT II)*, 932 F.2d 1504 (1991), *cert. denied*, 503 U.S. 913 (1992).

28 *Id.* at 1509.

29 Pub. L. 102-356, § 16, 47 U.S.C. § 303 note.

30 *In the Matter of Enforcement of Prohibitions Against Broadcast Indecency in 18 U.S.C. 1464*, 8 F.C.C. Rcd. 704 (1993).

31 *Action for Children's Television v. Federal Communications Commission*, 11 F.3d 170 (D.C. Cir. 1993).

32 8 F.C.C. Rcd. at 705 - 706.

33 11 F.3d at 171.

34 *Id.*

35 *Action for Children's Television v. Federal Communications Commission*, 15 F.3d 186 (D.C. Cir. 1994).

36 *Action for Children's Television v. Federal Communications Commission (ACT III)*, 58 F.3d 654 (D.C. Cir. 1995), *cert. denied*, 516 U.S. 1043 (1996).

37 58 F.3d at 656.

38 *Enforcement of Prohibitions Against Broadcast Indecency in 18 U.S.C. § 1464*, 10 F.C.C. Rcd. 10558 (1995); 47 C.F.R. 73.3999(b). Subsection (b) prohibits the broadcast of material which is obscene without any reference to time of day. Broadcast obscenity will not be discussed in this report. For more information on obscenity, see CRS Report 95-804, *Obscenity and Indecency: Constitutional Principles and Federal Statutes*, by Henry Cohen, and CRS Report 98-670, *Obscenity, Child Pornography, and Indecency: Recent Developments and Pending Issues*, by Henry Cohen.

39 60 FR 44439 (August 28, 1995).

40 Enforcement actions based on televised broadcast indecency are rare. However, the Commission recently issued a *Notice of Apparent Liability* for the broadcast of indecent material during a televised morning news program. During the program, the show's hosts interviewed performers with a production entitled "Puppetry of the Penis," who appeared wearing capes but were otherwise nude. A performer's penis was exposed during the broadcast. *See In the Matter of Young Broadcasting of San Francisco, Inc.*, File No. EB-02- IH-0786 (January 27, 2004). *See also In the Matter of Complaints Against Various Licensees Regarding Their Broadcast of the Fox Television Network Program "Married by America" on April 7, 2003*, File No. EB-03-IH-0162 (October 12, 2004).

41 According to statute, the maximum fine per violation is $25,000. *See* 47 U.S.C. 503(b)(2)(A). However, the maximum forfeiture amount was increased to $32,500 pursuant to the Debt Collection Improvement Act of 1996, Public Law 104-134, which

amends the Federal Civil Monetary Penalty Inflation Adjustment Act of 1990, Public Law 101-410. *See* 47 C.F.R. 1.80.

42 *See* fn 93, *infra* and accompanying text. Regulations set a maximum forfeiture of $325,000 for any single act or failure to act, which arguably limits the forfeiture for a single broadcast. *See* 47 C.F.R. 1.80.

43 *See In the Matter of Industry Guidance on the Commission's Case Law Interpreting 18 U.S.C. § 1464 and Enforcement Policies Regarding Broadcast Indecency*, File No. EB-00- IH-0089 (April 6, 2001). [http://www.fcc.gov/eb/Orders/2001/fcc01090.html]

44 The Commission's Enforcement Bureau initially dismissed the complaint against broadcast licensees for airing the 2003 Golden Globe Awards, in which performer Bono uttered the phrase "this is really, really f[***]ing brilliant," due primarily to the fact that the language in question did not describe or depict sexual and excretory activities or organs. 18 F.C.C. Rcd. 19859 (2003). The decision of the Enforcement Bureau has since been reversed. *See infra* regarding this case.

45 *Id.*

46 *Id.*

47 *Id.* at 5.

48 *Notice of Apparent Liability, State University of New York*, 8 F.C.C. Rcd. 456 (1993).

49 *Id.*

50 *See* note 35, *supra.*

51 *Notice of Apparent Liability, KGB Incorporated*, 7 F.C.C. Rcd. 3207 (1992). *See also Great American Television and Radio Company, Inc.*, 6 F.C.C. Rcd. 3692 (1990); *WIOD, Inc.*, 6 F.C.C. Rcd. 3704 (1989).

52 6 F.C.C. Rcd. 3692.

53 *See* note 35, *supra.*

54 *Notice of Apparent Liability, Citicasters Co.*, 13 F.C.C. Rcd. 22004 (1998).

55 The Commission has recently indicated that "the mere fact that specific words or phrases are not sustained or repeated does not mandate a finding that material that is otherwise patently offensive to the broadcast medium is not indecent." *In the Matter of Complaints Against Various Broadcast Licensees Regarding the Airing of the "Golden Globe Awards" Program*, File No. EB-03-IH-0110 (March 18, 2004). *See* section entitled Golden Globe (continued...) Awards Decision *infra.*

56 *L.M. Communications of South Carolina, Inc.*, 7 F.C.C. Rcd. 1595 (1992).

57 *Id.*

58 *See Notice of Apparent Liability, Temple Radio, Inc.*, 12 F.C.C. Rcd. 21828 (1997); *Notice of Apparent Liability, EZ New Orleans, Inc.*, 12 F.C.C. Rcd. 4147 (1997).

59 *Notice of Apparent Liability, Rusk Corporation, Radio Station KLOL*, 5 F.C.C. Rcd. 6332 (1990).

60 *In the Matter of Application for Review of the Dismissal of an Indecency Complaint Against King Broadcasting Co.*, 5 F.C.C. Rcd. 2971 (1990).

61 *Id.*

62 *Id.*

63 The Commission declined to impose a forfeiture on the broadcast licensees named in the complaint because they were not "on notice" regarding the new interpretations of the Commission's regulations regarding broadcast indecency and the newly adopted

definition of profanity. The Commission also indicated that it will not use its decision in this case adversely against the licensees during the license renewal process.

64 *In the Matter of Complaints Against Various Broadcast Licensees Regarding Their Airing of the "Golden Globe Awards" Program*, File No. EB-03-IH-0110 at 4 (March 18, 2004).

65 *Id.*

66 *Id.* at 5.

67 *Id.* at 6. *See* section entitled Dwelling or Repetition of Potentially Indecent Material *supra.*

68 *Id.*

69 *Id.* at 7. It should be noted that, although in this case the Commission found that the broadcast in question was both indecent and profane, there are certain to be words that could be deemed "profane," but do not fit the Commission's definition of "indecent." Under the newly adopted definition of "profanity," many words could arguably be found "profane" because they provoke "violent resentment" or are otherwise "grossly offensive," but not be found "indecent" because they do not refer to any sexual or excretory activity or organ or even "inherently" have a sexual connotation, as the Commission found the phrase that Bono uttered to have. Presumably, it is these words that the Commission will consider on a caseby- case basis.

70 *Id.* at 7, citing Black's Law Dictionary 1210 (6th ed. 1990) and American Heritage College Dictionary 1112 (4th ed. 2002).

71 *Id.*, citing *Tallman v. United States*, 465 F.2d 282, 286 (7th Cir. 1972).

72 *Id.*

73 *Complaints Against Various Television Licensees Concerning Their February 1, 2004, Broadcast of the Super Bowl XXXVIII Halftime Show*, File No. EB-04-IH-0011 (September 22, 2004).

74 *Id.* at ¶ 11.

75 *Id.* at ¶ 13.

76 *Id.* at ¶ 14.

77 *Id.*

78 *Id.* at ¶¶ 17 - 24.

79 47 C.F.R. 1.80(f)(3).

80 47 C.F.R. 1.80(f)(4).

81 Procedures for appealing Commission orders are set forth in 47 U.S.C. 402.

82 For a complete list of recent actions related to broadcast indecency, see [http://www.fcc.gov/eb/broadcast/obscind.html].

83 *In the Matter of Infinity Broadcasting, et al.*, EB-02-IH-0685 (October 2, 2003).

84 *Id.* at 8.

85 *Id.* at 9. The Commission noted that the contest portion of the broadcast in question lasted over an hour and was reproduced in a 203-page transcript.

86 *See In the Matter of Viacom Inc., Infinity Radio Inc., et. al.*, FCC 04-268 (November 23, 2004). [http://hraunfoss.fcc.gov/edocs_public/attachmatch/FCC-04-268A1.pdf]. The decree also covers several other actions pending against Viacom-owned Infinity Radio stations and broadcast television stations, but does not cover the proceedings related to the Super Bowl halftime show discussed *supra.*

87 *In the Matter of Clear Channel Broadcasting Licenses, Inc., et al.*, File No. EB-02-IH-0261 (January 27, 2004).

88 *Id.* at 4.

89 *Id.* at 5.

90 *Id.* at 6.

91 *Id.*

92 *In the Matter of Clear Channel Broadcasting Licensees*, File No. EB-03-IH — 159 (April 8, 2004).

93 *See* Statement of Commissioner Michael J. Copps, [http://hraunfoss.fcc.gov/edocs_public/attachmatch/DOC-245911A1.pdf], p. 2.

94 *See In the Matter of Clear Channel Communications, Inc.*, FCC 04-128 (June 9, 2004) at [http://hraunfoss.fcc.gov/edocs_public/attachmatch/FCC-04-128A1.pdf].

95 *Id.* at 7.

96 H.R. 3687. The FCC's indecency regulation prescribes the 10 p.m.-to-6 a.m. "safe harbor" for "indecent programming." No comparable regulation addresses profane programming. One might argue, therefore, that the ban on profane programming under 18 U.S.C. § 1464, as it exists now and as it would be amended by H.R. 3687, would apply around the clock. This reading of the statute, however, might render it unconstitutional (*see* note 27, *supra*), and a court might reject it for that reason. *See*, United States v. X-Citement Video, Inc., 513 U.S. 64, 69 (1994) ("a statute is to be construed where fairly possible so as to avoid substantial constitutional questions").

97 H.Rept. 108-434.

98 Roll Call No. 55, 150 Cong. Rec. H1035 (daily ed. March 11, 2004).

99 S.Rept. 108-253.

100 The Senate committee also added a new title to the original bill, entitled the "Children's Protection from Violent Programming Act."

101 S.Amdt. 3235, as amended.

102 *See* conference report for H.R. 4200, H.Rept. 108-767.

103 This bill appears to be substantially similar to H.R. 3717, as reported by the House Committee on Energy and Commerce, on March 3, 2004.

104 H.Amdt. 10. The FCC's policy statement on broadcast indecency was released on April 6, 2001, and can be found here [http://www.fcc.gov/eb/Orders/2001/fcc01090.html].

105 438 U.S. 726 (1978). The FCC's action was to issue "a declaratory order granting the complaint," and "state that the order would be 'associated with the station's license file,'" which means that the FCC could consider it when it came time for the station's license renewal. *Id.* at 730.

106 521 U.S. 844, 868 (1997) (noting that "the history of the extensive regulation of the broadcast medium" and "the scarcity of available frequencies" are factors "not present in cyberspace," and striking down parts of the Communications Decency Act of 1996). The Court also cited *Pacifica* with approval in *United States v. Playboy Entertainment Group, Inc.*, 529 U.S. 803, 813-814 (2000), and in *Ashcroft v. Free Speech Coalition*, 535 U.S. 234, 245 (2002).

107 *Pacifica, supra*, 438 U.S. at 750. A federal court of appeals subsequently held unconstitutional a federal statute that banned "indecent" broadcasts 24 hours a day, but, in a later case, the same court upheld the present statute, 47 U.S.C. § 303 note, which bans "indecent" broadcasts from 6 a.m. to 10 p.m. Action for Children's Television v.

FCC, 932 F.2d 1504 (D.C. Cir. 1991), *cert. denied*, 503 U.S. 913 (1992); Action for Children's Television v. FCC, 58 F.3d 654 (D.C. Cir. 1995) (en banc), *cert. denied*, 516 U.S. 1043 (1996).

108 *Id.* at 750, n.29.

109 *Id.* at 746 n.22. These two sentences and the text accompanying the next footnote, although part of Justice Stevens' opinion, are in a part of the opinion (IV-B) joined by only two other justices. Every other quotation from *Pacifica* in this report was from a part of the opinion that a majority of the justices joined.

110 *Id.* at 746.

111 There also appears to be some tension between, on the one hand, Justice Stevens' distinction in *Pacifica* between a point of view and the way in which it is expressed, and, on the other hand, the Court's statement in *Cohen v. California* "that much linguistic expression serves a dual communicative function: it conveys not only ideas capable of relatively precise, detached explication, but otherwise inexpressible emotions as well. In fact, words are often chosen as much for their emotive as their cognitive force. We cannot sanction the view that the Constitution, while solicitous of the cognitive content of individual speech, has little or no regard for that emotive function which, practically speaking, may often be the more important element of the overall message sought to be communicated." 403 U.S. 15, 26 (1971) (upholding the First Amendment right, in the corridor of a courthouse, to wear a jacket bearing the words "F[***] the Draft"). Arguably, Carlin's use of "indecent" words not only served an emotive purpose, but served to indicate the precise words to whose censorship he was objecting. Yet *Pacifica* was decided after *Cohen*, which suggests that *Cohen* does not lessen the precedential value of *Pacifica.*

112 Miller v. California, 413 U.S. 15, 27, 24 (1973). In addition, in striking down parts of the Communications Decency Act of 1996, the Court expressed concern that the statute may "extend to discussions about prison rape or safe sexual practices, artistic images that include nude subjects, and arguably the card catalogue of the Carnegie Library." *Reno v. ACLU, supra*, 521 U.S. at 878. And, in striking down a federal statute that prohibited child pornography that was produced without the use of an actual child, the Court expressed concern that the statute "prohibits speech despite its serious literary, artistic, political, or scientific value." Ashcroft v. Free Speech Coalition, 535 U.S. 234, 246 (2002). In neither of these cases, however, did the Court state that its holding turned on the statute's application to works of serious value.

113 390 U.S. 629 (1968).

114 *Pacifica, supra*, 438 U.S. at 731-732.

115 *Id.* at 732 (distinguishing "indecent" from "obscene" and "profane" in 18 U.S.C. § 1464).

116 *Id.* at 740.

117 *Id.* at 748-750.

118 395 U.S. 367, 369 (1969).

119 *Id.* at 388.

120 *Id.* at 394.

121 *Id.* at 396.

122 512 U.S. 622, 639 (1994). In *Turner*, the Court held that the "must carry" rules, which "require cable television systems to devote a portion of their channels to the transmission of local broadcast television stations," *id.* at 626, were content-neutral and therefore not

subject to strict scrutiny. The Court remanded and ultimately upheld the rules. Turner Broadcasting System, Inc., 520 U.S. 180 (1997).

123 In the court of appeals decision upholding the current statute that bans "indecent" broadcasts from 6 a.m. to 10 p.m., a dissenting judge wrote of "the utterly irrational distinction that Congress has created between *broadcast* and *cable* operators. No one disputes that cable exhibits more and worse indecency than does broadcast. And cable television is certainly pervasive in our country." Action for Children's Television v. FCC, *supra*, 58 F.3d at 671 (emphasis in original) (Edwards, C.J., dissenting).

124 512 U.S. at 638 (citation omitted).

125 Syracuse Peace Council v. FCC, 867 F.2d 654 (D.C. Cir. 1989), *cert. denied*, 493 U.S. 1019 (1990).

126 *Id.* at 669. In *Arkansas AFL-CIO v. FCC*, 11 F.3d 1430 (8th Cir. 1993) (en banc), the court of appeals held that Congress had not codified the fairness doctrine and that the FCC's decision to eliminate it was a reasonable interpretation of the statutory requirement that licensees operate in the public interest.

127 518 U.S. 727, 748 (1996). The plurality added that cable television "is as 'accessible to children' as over-the-air broadcasting, if not more so," has also "established a uniquely pervasive presence in the lives of all Americans," and can also "'confron[t] the citizen' in 'the privacy of the home,' ... with little or no prior warning." *Id.* at 744-745. Justice Souter concurred that "today's plurality opinion rightly observes that the characteristics of broadcast radio that rendered indecency particularly threatening in *Pacifica*, that is, its intrusion into the house and accessibility to children, are also present in the case of cable television. ..." *Id.* at 776.

128 529 U.S. 803, 813 (2000) (striking down a federal statute that required distributors to fully scramble or fully block signal bleed to non-subscribers to cable channels; "signal bleed" refers to the audio or visual portions of cable television programs that non-subscribers to a cable channel may be able to hear or see despite the fact that the programs have been scrambled to prevent the non-subscribers from hearing or seeing them).

129 An earlier district court case held that *Pacifica* does not apply to cable television because of several differences between cable and broadcasting. For one, "[i]n the cable medium, the physical scarcity that justifies content regulation in broadcasting is not present." For another, as a subscriber medium, "cable TV is not an intruder but an invitee whose invitation can be carefully circumscribed." Community Television v. Wilkinson, 611 F. Supp. 1099 (D. Utah 1985), *aff'd*, 800 F.2d 989 (10th Cir. 1986), *aff'd*, 480 U.S. 926 (1987) (striking down Utah Cable Television Programming Decency Act). The court of appeals did not discuss the constitutional issue beyond stating that it agreed with the district court's reasons for its holding. 800 F.2d at 991. A summary affirmance by the Supreme Court, as in this case, is "an affirmance of the judgment only," and does not indicate approval of the reasoning of the court below. Mandel v. Bradley, 432 U.S. 173, 176 (1977).

130 See CRS Report RL33170, *Constitutionality of Applying the FCC's Indecency Restriction to Cable Television*, by Henry Cohen, which concludes that "it appears likely that a court would find that to apply the FCC's indecency restriction to cable television would be unconstitutional."

131 Sable Communications of California v. Federal Communications Commission, 492 U.S. 115 (1989); Action for Children's Television v. FCC, *supra*, 932 F.2d at 1509.

132 *Id.* at 126.

133 Cohen v. California, *supra*, 403 U.S. at 19, 21. Under *Pacifica*, broadcast media do thrust words upon unwilling or unsuspecting viewers, but, if a court were to apply strict scrutiny to a ban on the broadcast of "indecent" language regardless of context, then it would not be following *Pacifica*.

134 *Turner Broadcasting*, *supra*, 512 U.S. at 664 (incidental restriction on speech). *See also*, Edenfield v. Fane, 507 U.S. 761, 770-771 (1993) (restriction on commercial speech); Nixon v. Shrink Missouri Government PAC, 528 U.S. 377, 392 (2000) (restriction on campaign contributions). In all three of these cases, the government had restricted less-than-fully protected speech, so the Court did not apply strict scrutiny. Because offensive words are apparently entitled to full First Amendment protection (except under *Pacifica* and in the instances cited in *Cohen v. California*, quoted in the text above), it seems all the more likely that the Court, if it applied strict scrutiny instead of *Pacifica* to a challenge to a ban on the broadcast of "indecent" words regardless of context, would require the government to demonstrate that harms it recites are real and that the ban would alleviate these harms in a direct and material way.

135 Interactive Digital Software Association v. St. Louis County, Missouri, 329 F.3d 954, 959 (8 th Cir. 2003). The Supreme Court has "recognized that there is a compelling interest in protecting the physical and psychological well-being of minors. This interest extends to shielding minors from the influence of literature that is not obscene by adult standards." *Sable*, *supra*, 492 U.S. at 126. The Court has also upheld a state law banning the distribution to minors of "so-called 'girlie' magazines" even as it acknowledged that "[i]t is very doubtful that this finding [that such magazines are "a basic factor in impairing the ethical and moral development of our youth"] expresses an accepted scientific fact." Ginsberg v. New York, *supra*, 390 U.S. at 631, 641. "To sustain state power to exclude [such material from minors]," the Court wrote, "requires only that we be able to say that it was not irrational for the legislature to find that exposure to material condemned by the statute is harmful to minors." *Id.* at 641. *Ginsberg* thus "invokes the much less exacting 'rational basis' standard of review," rather than strict scrutiny. *Interactive Digital Software Association*, *supra*, 329 F.3d at 959. A federal district court wrote: We are troubled by the absence of evidence of harm presented both before Congress and before us that the viewing of signal bleed of sexually explicit programming causes harm to children and that the avoidance of this harm can be recognized as a compelling State interest. We recognize that the Supreme Court's jurisprudence does not require empirical evidence. Only some minimal amount of evidence is required when sexually explicit programming and children are involved. Playboy Entertainment Group, Inc. v. United States, 30 F. Supp.2d 702, 716 (D. Del. 1998), *aff'd*, 529 U.S. 803 (2000). The district court therefore found that the statute served a compelling governmental interest, though it held it unconstitutional because it found that the statute did not constitute the least restrictive means to advance the interest. The Supreme Court affirmed on the same ground, apparently assuming the existence of a compelling governmental interest, but finding a less restrictive means that could have been used. In another case, a federal court of appeals, upholding the current statute that bans "indecent" broadcasts from 6 a.m. to 10 p.m., noted "that the Supreme Court has recognized that the

Government's interest in protecting children extends beyond shielding them from physical and psychological harm. The statute that the Court found constitutional in *Ginsberg* sought to protect children from exposure to materials that would 'impair [their] *ethical and moral* development. ... Congress does not need the testimony of psychiatrists and social scientists in order to take note of the coarsening of impressionable minds that can result from a persistent exposure to sexually explicit material. . . .'" Action for Children's Television v. FCC, *supra,* 58 F.3d at 662 (brackets and italics supplied by the court). A dissenting judge in the case noted that, "[t]here is not one iota of evidence in the record ... to support the claim that exposure to indecency is harmful — indeed, the nature of the alleged 'harm' is never explained." *Id.* at 671 (D.C. Cir. 1995) (Edwards, C.J., dissenting).

136 The full Commission's decision in the Bono case stated that "any use of that word or a variation, in any context, inherently has a sexual connotation." But this does not necessarily mean that it is sexually oriented enough to cause the courts to assume without evidence that it is harmful to minors.

137 *Ginsberg, supra,* 390 at 640, 639. *See also,* Action for Children's Television v. FCC, *supra,* 58 F.3d at 661.

138 Action for Children's Television v. FCC, *supra,* 58 F.3d at 670 (emphasis in original) (Edwards, C.J., dissenting).

139 *Pacifica, supra,* 438 U.S. at 732.

In: Advances in Communications and Media Research, Volume 4 ISBN 1-60021-189-5
Editor: William H. Miller, pp. 103-127

Chapter 6

MANAGING AND EXPLOITING INTERRUPTION IN MULTIPARTY TALK*

***Jakob Cromdal*† *and Karin Osvaldsson*‡**
Linköping University, Linkoping, Sweden

ABSTRACT

Competing for access to the conversational floor is sometimes thought of as a way of 'accomplishing power' in interaction. However, research on 'interactional power' easily becomes ideological to the degree where analysts risk losing sight of the very phenomena they claim to investigate. A symptom of this within studies of discourse organization is the relatively free traffic between concepts such as overlap, turn competition and interruption. Even more deceptive is the acclaimed – yet seldom thoroughly examined – relation of these conversational phenomena to dominance and power, both of which are often treated as discursive consequences of participants' gender or social status (e.g., Bargiella-Chiappini & Harris, 1996; Berger, 1994; Ferguson, 1977; Huls, 2000; Palmer, 1989; West & Zimmerman, 1983). In the present paper, we will leave the latter problems aside, focusing mainly on issues of overlapping speech and interruption. Specifically, drawing upon conversation analytic work on interruptive procedures (e.g., Bilmes, 1997; Hutchby, 1993), we will present a couple of instances of interruption occurring in disparate multiparty settings. Specifically, we will argue that the work of 'being interrupted' – a defining feature of interruptions in CA studies – may be accomplished by other parties than the interruptee, and also that participants may engage in such work to achieve local interactional goals. We begin this chapter with a brief presentation of some

* This article was prepared during our stay as visiting research fellows at the Department of Social Sciences, Loughborough University, financed by grants from the Swedish Foundation for International Cooperation in Research and Higher Education (STINT) and the Swedish Council for Working Life and Social Research (FAS). A portion of the data in this study were presented for the Discourse And Rhetoric Group (DARG), and we are indebted to more people than can be mentioned here for generous suggestions as how to proceed with the analysis.
† jakcr@tema.liu.se
‡ karos@tema.liu.se

issues of conversational organization relevant for our understanding of interruption in talk.

Some Introductory Issues on Conversational Organization

'One Speaker at a Time'

In the most general sense, research within conversation anslysis (henceforth CA) aims to explicate the generic methods through which participants are able to accomplish accountable actions within talk-in-interaction. Here, analyses of talk-in-interaction typically rely on the turn-taking machinery developed by Sacks (1992) and Sacks et al. (1974), to understand how utterances and other forms of action are deployed and coordinated within the flow of interaction. According to this account, the most general feature of conversational organization is that talk is systematically ordered so as to allow one party at a time to hold the conversational floor. However, this should not be taken to imply that simultaneous, or 'overlapping', talk can not occur within conversational exchanges. On the contrary, simultaneous talk is a frequent feature of conversational exchanges and a range of CA studies have focused on how participants deal with overlapping speech as part and parcel of talk's ordinary business. In so doing, however, participants typically orient in various ways to the underlying expectation that one party should talk at any one time (e.g., Hopper, 1992; Jefferson, 1986; Lerner, 1989; Sacks et al., 1974; Schegloff, 1987). Hence, the normative order of 'one party at a time' is overwhelmingly oriented to (but not necessarily adhered to at all times) as a 'seen but unnoticed' feature (Garfinkel, 1967) of participants' production of mutually recognizable actions.

Local Management of Overlapping Talk

In order to understand conversational interruption as a 'participants' phenomenon', recognized and oriented to as relevant for the trajectory of actions being pursued, we need to distinguish it from a variety of other forms of simultaneous speech. These issues are discussed in Drummond's (1989) useful review of studies on interruption, in which he stresses the need for analysts to attend to the details of overlapping speech.

A range of CA studies have explored the structural organization of overlapping passages. For instance Jefferson (1983) identified several forms of *overlap onset*. Briefly, the incoming speaker may begin talking upon recognizing an upcoming slot relevant for speaker transition ('transitional onset'), or even a perceived completeness of the current speaker's turn ('recognitional onset'). Now, these may be problematic issues for the participants involved, since there is no way of knowing, for the incoming speaker, whether the current speaker actually intended to give up the current turn or to continue speaking, and this is precisely the sort of interactional work in which participants engage as talk's ordinary business. Whether the outcome of this work results in actual interruption or not is of course an empirical issue, and needs to be investigated on a local basis.

Relevant for such analysis is the issue of *overlap resolution*, and several CA studies highlight the methods participants may exploit to terminate an overlapping passage (Lerner, 1989; Schegloff, 1987, 2000). These studies show that participants overwhelmingly work towards a soon closure of the overlapping passage, which may be seen as an orientation towards a notion of 'one party at a time' as a general and normative order of conversational conduct.

Now, whereas in most cases overlapping talk is dealt with so as to reduce overlapping stretches and allow one speaker to talk at a time, participants clearly have the option to 'go competitive' in pursuing whatever interactional projects that may be at hand (Cromdal, 2001a; French & Local, 1983; Schegloff, 1987). In an early study, French & Local (1983) provide a detailed account of various prosody-based techniques, such as modifications of pitch, volume and speech tempo, through which incoming speakers may employ to disruptively claim the floor. More recently, Schegloff (2000) detailed a range of procedures comprising an *overlap resolution device*, through which participants may strive to secure the current turn for their own purposes. Specifically, Schegloff presented two distinct sets of resources that may be used to produce turns at talk in the face of potentially turn-competitive simultaneous talk: *hitches*, comprising short obstructions in the progression of a turn (such as cutoffs, repetitions and sound elongations) and *perturbations*, that is, observable prosodic changes in the articulation of a turn or its components. Additionally, he identified some relevant locations in the local trajectory of overlapping passages, in which these procedures may be deployed. In this procedural account of the interactional mechanisms underlying the situated workings of this overlap resolution device Schegloff argued for "an interactional 'logic' by which these resources, in those places, constitute 'moves' of a describable sort in a competitive sequential topography" (2000, p. 11).

Clearly then, from a members' perspective on social interaction, a disciplined examination of the local practices through which members handle simultaneous speech and which comprise the situated accomplishment of conversational interruption, should form part of any discussion of the power relations that may or may not be at stake for the participants themselves in conversational exchanges.

INTERRUPTION AS MEMBERS' PRACTICE: 'DOING INTERRUPTION' AND 'DOING BEING INTERRUPTED'

In commonsense terms, occurrences of simultaneous talk are frequently seen as equivalent with interruption. For instance, if we witness a speaker starting to talk in the midst of another participant's speech, then that observation is often thought of, in lay terms, as an interruption. In this sense, interruption -as a vernacular category- has an objective sense to it (Bilmes, 1997). Indeed, many previous studies occupied with issues of power and dominance in communicative exchanges have appealed to this notion of interruption (e.g., Bargiella-Chiappini & Harris, 1996; Huls, 2000; Rogers & Jones, 1980; West, 1979).

In the previous sections, we have argued that there is good reason to refine this practice, by attending to the minute organizational details of the situated practices, through which conversational interruptions are accomplished. Moreover, although a rigorous account of onset and resolution procedures may be seen as an 'optimal starting point from which to make claims about conversational dominance' (Drummond, 1989, p. 151), we would suggest that analyses of the formal aspects of overlapping talk and conversationally disruptive phenomena also need to attend to and make explicit the interactional upshots of such structural phenomena for the participants themselves, that is for their interactional business at hand. As Bennett (1981) points out, whereas 'overlap' can be seen as an observation of a structural feature of discourse, 'interruption' is clearly an interpretive analytic category. In a similar vein, Bilmes (1997) points out that interruption can be seen "as a normative phenomenon solely produced and recognized by participants" (p. 511, emphasis in original). This means that what constitutes an interruption for the participants at talk, can not be specified in advance according to a formal analytical scheme of overlap resolution, particularly since interruption may be seen to take place without any prior occurrence of overlap (Murray, 1985). Accordingly, Bilmes further asserts that "in order for an event to be an interruption, it is not enough that it meet some formal criteria; it is not enough that it be perceived as a violation of speaking rights by the analyst, by members, or even by the participants themselves. It must be *displayed and handled* as violative within the interaction" (1997, p. 512, emphasis in original). Hence, the issue of interruptive violation of speakers' rights must be understood as an issue for the participants themselves, rather than solely for the analyst (see also, Bennett, 1981; Eglin, 2000; Hutchby, 1993). Accordingly, a central task for the analyst is to show how participants at talk orient towards an incoming action as recognizably interruptive. In sum, analyses of conversational interruption must attend to both the interactants' ways of 'doing the interrupting' as well as to their ways of 'doing being interrupted' (Bilmes, 1997; Cromdal, 2001a).

Aims and Materials

As we have seen above, earlier work within conversation analysis has presented detailed accounts of such conversational phenomena as overlap and interruption. We should note however, that previous studies mainly deal with dyadic exchanges. However, as both overlapping talk and interruption are perfectly ordinary phenomena in multiparty settings, our aim is to extend the analytical scope to entail interaction taking place within larger groups of participants. For this reason, the present paper offers some preliminary considerations of samples of multiparty interactions. Specifically, the goals of the paper at hand are to:

- review previously identified structural constellations of overlapping exchanges and point to some ways these can be extended to include interaction in multiparty settings.
- demonstrate on the basis of detailed sequential analysis of interaction, that as a members' phenomenon, interruption may be at stake not just for the speakers whose talk is being overlapped, but also to the parties acting as recipients of this talk.
- show that, accordingly, an incoming turn may be treated as interruptive of ongoing talk by any party to an an interactional episode, and crucially, that such actions – through which interruptions are socially established- may be exploited for local interactional purposes.

To substantiate these claims we draw upon two samples of data recorded in very distinct types of social settings. The empirical example is taken from a corpus of videotaped recess interaction among bilingual children attending 2nd grade in an english school in Sweden (cf. Cromdal & Aronsson, 2000; Cromdal, 2001a, 2004). The second example is taken from a large multiparty event, a so-called network conference in a Swedish Youth Detention Home. Such conferences normally involve a number of participants typically representing three different parties: (i) the resident youth herself and one or several family members, (ii) representatives of the referring social services authorities, and (iii) staff from the Detention Home, representing various fields of expertise, mostly social workers, psychologists and teachers (Osvaldsson, 2002; 2004/in press, for more extensive presentations of this field).

MANAGING OVERLAP AND INTERRUPTION IN MULTIPARTY PLAY

Our analysis begins by briefly considering a stretch of multiparty interaction in a group of 2nd grade primary school children at play. This will lead us to focus on different patterns of overlapping talk and to discuss, in more detail, some features of conversational interruption. In the second part of the analysis, we present an empirical extract from a multiparty meeting at a youth detention home, which will allow us to further discuss a few conversational features that are not only interruptive with respect to the locally unfolding interaction, but which are recognizably disruptive of the institutional speaking order that has been established for the purposes of this meeting.

In the first extract, the children are involved in a popular recess activity we have previously labeled *koja*-play.[1] The transcript begins over 10 minutes into the current play episode, during which a number of aggressors have attempted to raid the koja. Likewise, several boys (including Gustav and William in the transcript below) have been attempting to gain entry into the koja by peaceful means, for instance by volunteering enrollment as guards.

In rough outline, the transcript starts with William being tricked by Beatrice into believing that he is accepted on the koja-team,[2] just to find out in line 3 that this will not grant him entry into the koja. His triumphant exclamation in line 2 is soon transformed into a display of strong disappointment in line 4 and followed by collective laughter (lines 3, 5 and 6) by the girls on the koja-team, through which they orient to (and arguably enjoy) the

---

1 The Swedish term 'koja' translates to English as 'hut', 'shack', or even 'treehouse'. When referring to children's hideouts, all participants in the study (children and staff) used the Swedish term. Hence, in conversations conducted in English, *koja* was used as a borrowed lexical unit. For this reason, we will stick to this term throughout the analysis.

Koja play was the most frequent type of group activity during recess at the school. A play episode would begin with a group of children designating some suitable location as their hideout (behind some trees, up in a tree, or as in the present case, up on a shelf in the rocks). Virtually as soon as the team has settled in their koja – and sometimes even sooner – another team is formed, with the sole purpose of raiding, plundering and sometimes pillaging the koja.

2 The arrangement of koja teams often opens various possibilities for current non-players to attempt enrollment into one of the two teams. Often, boys will attempt to enroll as guards for the *koja*. As I have shown elswhere (Cromdal, 2001b), the standard verbal bid for play entry (regardless of the type of play) is 'får jag va med?', or 'can I be with?' in English exchanges, which is the literal translation of the Swedish routine. Notably, while perfectly nonidiomatic in any standard variety of English, this routine was used by all children in the class, and sometimes even the (English-born) teachers: 'were you with, playing?' or 'did you ask if you could be with?'. For this reason, 'being with' it may be seen as a borrowed phrasal construction, established as part of the local variety of English spoken at the school.

successful deception; a trick well played. In line 9 however, Beatrice's decision not to let William in to the koja is opposed by Magdalena, who, exaggerating a pleading voice, tries to convince her to reconsider: 'oh ↓yes he ca:n'. Picking up the format of Magdalena's plea, albeit in a much more assertive tone of voice, Beatrice declines the request: 'oh no he *ca*:::n't'. Upon this follows a series of contradictory claims, in which the two girls (as well as Barbara in line 15) construct and escalate their oppositional stances (Cromdal, 2004; Goodwin, 1998) by exploiting the structural properties of the immediately preceeding turn, a conversational technique Goodwin (1990) labelled 'format tying'.

Excerpt 1. Continuing play in the koja. Participants: *Mag*dalena; *Bea*trice; *Will*iam; *Jola*nta; *Barb*ara, *Ste*ven, *Gus*tav and *Van*. [yr2p1t1a 2924-3006]

Bea: ok William you'e wi:th
Will: YEA:H
Bea: but you're NOT comin in the koja[(.7)] he[hhehaha
Will: [o↓↓::h]
Mag: [höhehh
Barb: [hehhehe
Ste: ((to Van at koja entrance)) dunno how to ( ) (if it)
works like *thi*s
Mag: ((to Beatrice, pleading voice)) oh ↓yes he ca:n.
Bea: oh no he *ca*:: [:n't.]
Mag: [oh y]es he *ca*::n.
Bea: oh no he *c* [*a*:::[n't
Gus: [WILL[IAM CAN I TALK]TO YOU.
Mag: [yes [he ca::]n.]
Barb: [yes he can.]
Bea: oh *no*=
Mag: =*yes*=
Bea: =no:
Mag: ye:s
Barb: ((facing everyone down below the koja)) hello:!
it's [*A:R ES: TE*::!!]= ((spelling 'R S T'))
Gus: [>william can I talk to you<]?
Bea: =((at barb)) SHH! SHH! SH [H!
Gus: [>william can
I [talk (to you?) (.) I ha]ve-[(I think)]
Bea: [((to all)) NOT ANY*MORE*!] [*cha*nge of pass]words!
Gus: I think [-
Jola: [WHO WANTS TO *KILL*! WHO WANTS TO *KILL*!
WHOh WhANTS TO *KhILL*! e↑heh= ((chanting))
Barb: =>c'm here< (.) [everybody come]
Gus: [hallo William][1] I' [(ve been up there)]
Van: ((to Steven)) [*yo*u're not allowed]
to be guard=

34. Gus: =and this pla [ce is ( ) [( )]
35. Barb: [EVERYBODY LET [( )]
36. Mag: [*hal*]*l*↑*å:GU*[2]*::STAV* (.)
hello: GU::STAV (.)
37. >*ly* [*ssna pa mej*< (*först*)]=
>*listen to me*< (*first*)=
38. Barb: [A N Y B O D Y *WI*:TH?]=
39. Gus: =*hey* [3]I tell ya I just [[4]think it's- that we just ( )
40. Ste: ((to Mag)) [HA*LLO* (VAN) SHOT A STICK AT ME=
41. Barb: =SH↑↑*U:T* [(.)ev [ery]body]
42. Gus: [just tell th[em to( )] ((to Will))
43. Will: ((facing Steven)) [S H ↑ U]D U]::P=
44. Gus: =>just tell[5] them< to ( ) like when we [were in ( ])
45. Mag: ((shouting at Gustav's face)) [CUT IT *OU*:T]
46. ((general uproar follows as Bud, the leader of the opposing
47. team, approaches the koja))

The unfolding impasse is finally resolved by Barbara, who turns to the group of boys beneath the brink of the koja and starts divulging the password required for entry (lines 20-21).[3] She is immediately hushed by Beatrice (line 23), who goes on to announce in line 26 that the password is now expired and is about to be changed. In the following interaction, Barbara makes four attempts to solicit the joint attention of the other participants to this exchange (lines 30, 35, 38 and 41). We also find Magdalena attempting to disrupt the ongoing interaction between Gustav and William (lines 36-37), by switching into Swedish[4] and shouting at Gustav to pay attention to what she has to say, and in line 45 where she explicitly demants that Gustav terminate the exchange with William.

Hence, another feature to attend to in Excerpt 1 is Gustav's interaction with William. Here, we may note that Gustav makes repeated attempts to establish a joint interactional project with the other boy (lines 13, 22, 24-25 and 27). There is no way of telling from the videorecording whether any of these attempts to solicit Williams attention are successful, but there is little doubt that with his turn in lines 31 and 34, Gustav has managed to align William with his interactional project, which – for all its incoherence due to overlapping speech and general uproar – seems to aim at establishing a joint stance against the girls in the koja, possibly even trying to talk his friend out of any further efforts to join the koja team.

The above account of the events in Excerpt 1 deliberately omits several turns represented in the transcript which constitute parts of an exchange between Steven and Van (lines 7-8 and 32-33), as well as the chanting of Jolanta in lines 28-29. The reason for doing this is simply that, with an exception for Steven's exclamation in line 40, none of these interactional projects are oriented to by the participants engaged in the exchange between the three girls on the koja team on the one hand, and the two boys beneath the brink of the koja, on the other. Also, the interaction taking place in lines 36 through 45 deserves more attention, as it entails

[3] We are not able to tell whether this was indeed the true password of a false one. However, announcing a fake password to attract visitors who would then be rejected, was almost an institutionalized component of *koja* play.

[4] See Cromdal (2001a), for an extensive analysis of children's use of code-switching in turn-competitive sequential environments.

several instances of overlapping talk in, or indeed through, which the participants demonstrably compete for the conversational floor. Let us therefore proceed to consider, in some analytical detail, the different forms of overlap that occurred in this excerpt to explicate their relevance for our understanding of conversational interruption as a matter of participants' orientation and interactional work.

Overlap Configurations in Multiparty Talk

In his seminal account of members' practices of managing simultaneous talk, Schegloff (2000) identifies three types of structural constellations of overlap exchanges, as schematized below:

(i) A → B	(ii) A → B	(iii) A ↔ B
↓	↑	
C	C	[C]

In Schegloff's account, pattern (iii) may be seen as be the most prototypical one in that it simply involves two speakers who address each other simultaneously. Naturally, simultaneous talk will also occur in multiparty exchanges, and in (i) and (ii) Schegloff presents two triadic scenarios. In configuration (i) speaker A addresses speaker B, who simultaneously addresses participant C, whereas in (ii) participant B is being simultaneously addressed by speakers A and C. An example of configuration (i) may be found in the final lines of Excerpt 1 as shown below:

44. Gus: ((toC)) =>just tell[5] them< to ( ) like when we [were in ( ])
45. Mag: ((shouting at Gustav's face)) [CUT IT *O*U:T]

Here, Magdalena (A) places her command in the midst of Gustav's (B) turn which is addressed to William (C). Configurations (i) and (ii) are rather straightforward scenarios in triadic exchanges and Schegloff additionally points out that they 'can naturally alternate under the operation of the turn-taking system' (p. 8). That is to say, participants B and C may engage in taking turns at talk *in overlap with* speaker A.[5]

The configurations sketched out by Schegloff involve three participants or less. However, in conversational exchanges which involve more than three participants, we would expect to find other, more complex constellations. In particular, multiparty exchanges allow speakers to allocate turns to more than one interlocutor. We will offer one example to illustrate this possibility. The following example (lines 39-43 in Excerpt 1) can be preliminarily considered a variation on Schegloff's configuration (i):

---

5 Akin to this constallation is a scenario in which speakers B and C address each other simultaneously and in overlap with speaker A. This scenario would be schematized as follows:

A→B
↕
C

Note that this configuration does not occur in the data presented here –it is merely another logically derived possibility of simultaneous talk taking place among three participants.

39	Gus:	=*hey* ^{3}I tell ya I just	[^{4}think it's- that we just ( )
40	Ste:	((to Mag))	[HA*LLO* (VAN) SHOT A STICK AT ME=
→41	Barb:	=SH↑↑*U:T*	[(.)ev [ery]body]
→42	Gus:	[just tell th[em to( )] ((to Will))	
→43	Will:((facing Steven))		[S H ↑ U]D U]::P=

In line 42, Gustav is talking to William, who yells at Steven to shut up, in partial overlap with Gustav's turn. Hence, this stretch of simultaneous talk resembles a 'A talks to B, who talks to C' situation as described in Schegloff's configuration (i). However, such a characterization of the exchange assumes that the role of the fourth participant, Barbara (D), can be ignored. This can be analytically defensible provided that Barbara is clearly a non-ratified participant in this exchange, i.e., that her talk does not belong to the present exchange. In the present case, however, there are strong indications that Barbara's turn is in fact very much part of the interaction in lines 39-43. First off, it needs pointing out that the fact that the second part of her turn includes everyone present as possible recipients does not necessarily make her a ratified participant in the exchange, because none of the other participans seem to display any orientation to her attempts at securing the group's attention (lines 30; 35; 38 and 41).

There is however reason to believe that the first part of Barbara's turn in which she shouts at Steven to 'shut' (it/up?) is received and oriented to by William in line 43. First of all, we may note that they both respond to Steven's previous shouting with the same type of demand, and in that sense we are dealing with a type of incipient alignment, a common stance with respect to Steven's actions. Moreover, the beginning of William's turn is canonical with Barbara's with respect to such prosodic features as intonation, pitch, voice quality, amplitude and first vowel elongation. The last observation is particularly significant in showing that D's talk is consequential for the unfolding interaction. Hence, our analytical account of the overlapping talk in lines 41-43 will have to include Barbara's turn. This account can be schematized as follows:

A→ B
↑ ↗↓
D→ **C**
↓
X, Y, Z ...

Here, Schegloff's configuration (i) is extended with speaker D, who in addition to addressing participant C also addresses the other two speakers, as well as any other parties within earshot. This brief discussion of the configurations of simultaneous exchanges was to demonstrate that in multiparty activities, overlapping talk may be structurally more complex than in the dyadic or triadic constellations previously examined by conversation analysts. However, as Schegloff (2000) points out, there is also a need to find out whether participants' ways of handling overlapping talk in multiparty settings differ from the practices through which this is accomplished in dyadic exchanges and, if so, in what ways are those methods modified. Without going into analytical detail at this point, it would seem that whereas Barbara (D) simply drops out of the overlapping passage, Gustav (A) goes at lengths to persist in keeping his turn. We will discuss some of the methods through which he pursues this goal in the next section.

Who is being Interrupted?

Having observed that multiparty exchanges allow for somewhat complex structural constellations of simultaneous talk, we will now turn to consider how such exchanges are dealt with by the participants. Specifically, we will focus on two candidate instances of conversational interruption, and point to the means through which the participants work to establish a prior stretch of talk as interruptive of the currently ongoing interactional business.

We have already noted that in the present transcript Gustav is putting considerable effort into soliciting William's attention and into aligning with the other boy against the girls in the koja-team. We will now flesh out some features of the interaction taking place between the two boys, as they provide two illustrative cases of conversational interruption. Let us first take another look at the rather dense cluster of exchanges in lines 30 through 45:

30 Barb:	=>c'm here< (.) [everybody come]
31 Gus:	[hallo William][1] I' [(ve been up there)]
32 Van:	((to Steven)) [*yo*u're not allowed]
33	to be guard=
34 Gus:	=and this pla [ce is ( )[( )]
35 Barb:	[EVERYBODY LET [( )]
36 Mag:	[*hal*]*l*↑*å:GU*[2]*::STAV* (.)
	hello: GU::STAV (.)
37	>*ly* [*ssna pa mej*< (*först*)]=
	>*listen to me*< (*first*)=
38 Barb:	[A N Y B O D Y *WI*:TH?]=
39 Gus:	=*hey* [3]I tell ya I just [[4]think it's- that we just ( )
40 Ste:	((to Mag)) [HA*LLO* (VAN) SHOT A STICK AT ME=
41 Barb:	=SH↑↑*U:T* [(.)ev [ery]body]
42 Gus:	[just tell th[em to( )] ((to Will))
43 Will:	((facing Steven)) [S H ↑ U]D U]::P=
44 Gus:	=>just tell[5] them< to ( ) like when we [were in ( ])
45 Mag:	((shouting at Gustav's face)) [CUT IT *O*U:T]

We can see that in the interaction between Gustav and William it is Gustav who does the talking. Hence, in order to sustain the claim that there is in fact interaction taking place between the two boys, we need to show that William is at least orienting to Gustav; that he is demonstrably acting as recipient of Gustav's talk. We therefore begin by focusing on William's nonvocal moves throughout this part of the transcript. For reasons of transcription precision, these moves are indicated by numbers in superscript. Hence, at point 1 (line 31), William shifts his body posture so that he is observably orienting to Gustav, who has just made a new attempt to gain William's attention. To this end, Gustav made use of a typical attention-soliciting device, namely attention token + name ('hallo William'). At point 2 (line

36), William lifts his head to face Magdalena, who has just started to shout at Gustav. At point 3 (line 39), William briefly shifts attention back to Gustav in response to another attention token ('*hey*'). He then looks up at Steven (point 4; line 39), who has just approached the group, shouting at Magdalena (line 40). Having yelled back at Steven to shut up in line 43, William finally returns to attend to Gustav's talk at point 5 (line 44). In other words, although William can hardly be said to pay undivided attention to Gustav's talk, with several disruptions caused by other speakers, there should be no doubt that he does orient to the fact that Gustav is talking to him. Indeed, we intend to argue that Willian is in fact acting as one of the parties being interrupted in his interaction with Gustav.

But let us first take a look at the initial candidate ocurrence of interruption taking place in lines 34 through 36. Gustav's turn in line 34 is overlapped by two participants potentially bidding for his attention. First we have Barbara, who is trying to solicit the attention of all the group. In addition, we find Magdalena trying to secure the attention of an, at the point of the overlap, unnamed speaker.[6] Due to the heavy overlap in lines 34-36, it is not possible to tell whether Gustav drops out without completing his turn, or whether he finishes in competitive overlap with the two girls.[7] For this reason, it is impossible to claim that Gustav's talk *per se* has been subject to interruption. What has been disrupted by Magdalena's turn-competitive incoming (French & Local, 1983) in line 36, however, is William's attention to Gustav's talk. That this event is in fact treated as interruptive of the interaction between the two boys can be observed in Gustav's way of reclaiming Williams attention in line 39: '*hey* (.) I tell ya'. The use of the emphasized attention token and its minute timing with the projectable end of Barbara's turn in line 38 suggest that he still has something to tell William and that he therefore needs to regain his attention.

In other words, whether Magdalena's competitively incoming turn did or did not disrupt Gustav's actual production of talk, his way of reclaiming William's attention suggests that it certainly did disrupt the interaction between the two boys. This instance therefore offers itself to the analytical characterization as 'doing being interrupted', and we may argue that conversational interruption is not merely a matter of disrupting the speech production of the current speaker, but may also occur when the conversational activities of the participant(s) currently 'doing recipiency' are being disrupted or otherwise hampered by an incoming speaker.

Let us elaborate this point by considering a second candidate interruption, which takes place in lines 39-40. As we have seen above, line 39 begins with Gustav attempting to reclaim Williams attention, by placing an emphatically stressed attention-getting device ('*hey*') immediately after the projectable end of the prior turn. A micropause later he does receive William's attention, if only for a short time. In line 40, we can see Steven shouting to Magdalena that Van had apparently thrown a stick at him. Although his shouting does not appear to be part of the current exchange between Gustav and William, we can see that it

[6] It could of course be argued that there may be visual cues that allocate Magdalena's attention bid at Gustav right from the beginning of her turn in line 36. As far as we are able to tell from the videotape, however, from where she is standing, Magdalena could equally well be addressing William. It could actually be the case, that this is how William interprets Magdalena's turn to be allocated and therefore looks up to face her. However, since participants may orient to turns at talk whether they are addressed to them or not, this possibility remains purely speculative.

[7] This of course makes it impossible to tell whether William may have been able to hear (or project) the end of Gustav's turn before looking up at Magdalena, or whether he simply disattends Gustav's talk in favor for Magdalena's turn-competitive incoming in line 36.

starts in the midst of Gustav's talk. More importantly, we may note that it does seem consequential, in different ways, to the ongoing interaction.

First, Steven's shouting seems to attract William's attention, and we can see William looking up roughly at the same time as Steven begins to shout.[8] Second, since the shouting is clearly more audible than Gustav's talk, is poses a potential threat to the hearability of Gustav's talk. Either separately or together, these two observations may account for Gustav's cut off- reformulation in line 39 ('I just think it's- that we just'). It is important to note, however, that Steven's shouting does not prevent Gustav from continuing his talk, as in lines 37-38, but merely seems to affect the flow of Gustav's turn.[9] This illustrates Gustav's procedure for dealing with the overlap, thus revealing his orientation to the competitive and potentially disruptive features of Steven's shouting. Another participant who reacts to the shouting is Barbara who shouts back, demanding that Steven 'SH↑↑*U:T*' (up/it), thus treating Steven's turn as disruptive of her own interactional project: to solicit the attention of the entire group. Finally, we can see William responding to Steven's shouting by yelling at him to 'SH↑UD U::P', thus as we pointed out earlier, exploiting the format of Barbara's immediately preceeding turn. For our current purposes, we should note that through his response, William too orients to Steven's shouting as disruptive of his current interactional activity, that is, his recipiency of Gustav's talk. On the level of displayed action, we may thus gloss William's response as accomplishing 'being interrupted at listening to Gustav'.

The above instances both demonstrate how the recipients' actions are being disrupted by interjacent turns. It should be noted that in neither of the two cases was the interjacent talk actually addressed at the current recipient. In the first case, the interjacent talk was directed at the current speaker, and in the second instance the interjacent talk was directed at an interlocutor participating in an entirely different exchange. However, we need to bear in mind that these interjacent turns were not clearly allocated from the outset. Considering that the allocation of any turn must be handled interactively, as part of an instant negotiation between speaker and hearer -and in multiparty settings this may include an array of potential recipients- it is perhaps not very surprising to find the recipients orienting to the interjacent turns and by the same token, shifting attention from the talk to which they have been attending so far.

These observations lead us to propose a respecification of the very notion of conversational interruption. If we take interruption to be a dialogic phenomenon, we should not expect to find it only relevant to the actions of speakers. Accordingly, our analyses above show that it is not necessarlily the current speaker, whose actions are being overlapped by an interrupting party, that will orient to the event as an interruption. The interactional work of constructing incoming talk as disruptive of the current exchange may also be carried out by the receiving party, or parties.

In sum, when an interruption occurs, it is not just the speaker who is interrupted: it is the *interaction* between minimally two parties. In dyadic exchanges, however, this point is easily overlooked, since both parties are obviously engaged in speaking. It is by paying detailed analytical attention to multiparty interactions like the one we have seen above that we may observe this more clearly. This view maintains that conversational interruption is to be seen

---

[8] This is in fact very similar to what happens in line 36 in that both turns start off with a general, non-allocated bid for attention ('hallo').

as a members' phenomenon, rather that the analyst's, and we have demonstrated some of the means through which parties to an interaction go about 'doing being interrupted'. In the next section we will take the analysis one step further to consider how participants not only accomplish, but also exploit conversational interruptions for local interactional purposes.

OVERLAP AND INTERRUPTIVE PROCEDURES IN A MULTIPARTY MEETING

Our final extract will lead us to consider yet another instance of the practice we have termed 'doing being interrupted', and our analysis will highlight in some detail the local interactional upshots of this practice. In short, we intend to give some concrete proof of the type of actions that may be accomplished by claiming that an interruption has just occurred. The transcript is drawn from a corpus of network meetings taking place in a Swedish youth detention home (Osvaldsson, 2002). Such meetings are arranged prior to every assessment being conducted at the detention home, and the main purpose of these meetings is to arrive at a set of concrete questions that will guide the upcoming 8-week assessment of the young person's previous and current life situation. Such meetings are chaired by a senior member of the staff at the detention home, who begins by presenting the agenda. The referring authorities will then present their case, after which all participants get a chance to add to the list of issues for the assessment, or generally to comment on previous discussions. During this phase of the meeting, talking is typically organized 'tablewise', with chair making sure that all participants get an opportunity to talk.[10] One of the issues to be discussed in the analysis below deals with the participants orienting their talk towards this speaker agenda.

Just prior to this excerpt chair has finished writing down one of the social workers' earlier question on a whiteboard, and asks if there are any more issues that should be discussed. The transcript presents an lengthy interactional episode, which we will need to consider in brief before we turn to a more detailed examination of the interruptive sequential environment towards the end of the episode.

Excerpt 2. [AA 111] Present at the meeting were the following persons: Anita and her sister, Tina; two assistants (ass1, ass2) from the detention home; a family counsellor (fam); a psychologist (psy); as well as a senior psychologist who also chaired the meeting (chair). Representing the social welfare agencies are two social workers (soc1, soc2). Finally, the researcher (res) was present at the table, audiotaping the interaction and taking notes. The transcript begins approximately 30 minutes into the meeting

1	chair	me*:r*?
		mo*:re*?
2		(1)
3	soc1	mm ja den h↑*är* e lite speheciellheh
		mm well th↑is one is a bit spehehcialheh

---

9 On a speculative note, this may have to do with the fact that Magdalena's turn in lines 36-37 is specifically allocated at Gustav, wheres Steven's shouting does is not contain such an address.

10 This does not include the observing researcher.

4		(.5)
5		*o*m nu Anita
		*i*f now Anita
6		(.3)
7		kan tänka sej nån fo:rm av hjaelp så (.) har de kommit fram
		may agree to some fo:rm of help then (.) it has been found
8	Anita	mm
9	soc1	här nu >Anita< att m*ä*nskor som e sn↓älla dom [e n*ö:* *]rd*ar
		here now >Anita< that p*eo*ple who're n↓ice they' [re n*e:* *]rd*s
10	Anita	[mm]
11	soc1	(.) hur kan man hjälpa (.) *as*så: var kommer de:: fr*å:*n
		(.) how can you help (.) I'*me*:an where's tha::t come fr*o:*m
12		(2.6)
13	chair	*ja:*
		*ye:*ah
14	psy	va s*a* du ja uppf-(.)°in-° [alla som e snälla e
		what's th*a*t I didn'-(.)° n' °- [everyone nice 're
15		(?) [hhihiheheheheheheheheh
16	soc2	nördar
		nerds
17	(Anita)	nördar e:
		nerds 're:
18	Ass1	(e ja de då)
		(am I one then)
19	chair	tönt [ar ja:
		geek [s ri:ght
20	unison	[hehehehehehe[heh heh heh]ehehehehheheheh
21	chair	[↑*ja*: ja]
		[↑ri:ght right]
22	psy	okej
		okay
23	soc2	töntar
		geeks
24	soc1	men de blir- men de kan ju bli rätt så intress*an*t ju (.)
		but that'll be- that may turn out rather *in*teresting right(.)
25		om nu Anita (.2) känner att hon beh*ö*ver nån gång nåt
		should Anita (.2) feels that she somet*im*es needs some
26		stöd eller hjälp
		support or help
27		(.3)
28	Anita	(men-)
		(but-)
29	soc1	*hu*r man- [*hu*r man liksom kan göra de rätt [å så::
		*ho*w you- [*ho*w would you go about doin it like properly
	[ri::ght	
30	soc2	[( )
31	Anita	[>ja
		[>yeh

32		men får ja bara< säga *e*n grej, ja har nog *sa*gt de när
		but may I just< say *on*e thing, I probably *sa*id that when
33		jag har vatt väldigt↑*upp*↓rörd när jag har känt (.2)
		I was very ↑*up*↓*se*t when I felt (.2)
34		jag har väldigt (.5) >ja har< haft mycke probl*e*m (.5)
		that I've got very (.5) > I've< had a lotta pro*b*lems (.5)
35		>å då har de< bara hoppat ut ur *mu*nnen på mej. (inte sagt)=
		>an'so it's< just came right outta my *mou*th. (not said)=
36	chair	=så-
		=so-
37	Anita	så de är nog de hela va.
		so that's probably all there is to it right.
38	chair	ja ↑ja,
		right ↑right
39	Anita	att jag (.) har s*ä*kert varit (.) >besviken och *u*pprörd<
		that I've (.) surely been (.) >disappointed and ups*et*<
40		(.9)
41	?	°mm°
42		(.7)
43	soc1	men de har väl *va*rt (.4) visst en eller två som har vart (.)
		but hasn't there *be*en (.4) one or two surely who've been (.)
44		snälla å så va.
		nice an' all right.
45		(1.6)
46	soc1	°(om man ska hjäl-)°
		°(if you're to hel-)°
47	Anita	men de *är* så h↓*ä*r va (.) hhrrm ((clears throat))
		but it *wo*rks like th↓is right (.) errm ((clears throat))
48		om man *ha*r mycke problem och såna grejer man
		if you've *got* a lotta problems 'n stuff like that you
49		*u*ppfattar aldrig *snä*:llhet eller s↑*å* man (tro) å *se*n
		never *no*tice *ni:c*eness or ↑*an*ything (think) an' *the*n
50		när man har haft det som ja har *ha*ft de hela tiden (.)
		when you have *be*en where I've been all the time (.)
51		man kan inte se att nån är snäll mot en de [t]
		you're unable to notice any one being nice to ya yo [u]
52	(?)	[nä]
		[no]
53	Anita	GÅR inte (.)de e de som e de stora prob*lem*et (.5) men as↑så
		just CAN'T (.) that's the big *prob*lem (.5) but r↑ight
54		(.5) j↑a de e *m↑in* [(.3)] *å*sikt i de *he*la.
		(.5) w↑ell that's *m↑y* [(.3)] opinion 'bout the whole thing.
55	chair	[o:k↓ej]
		[o:k↓ay]
56	Anita	man[*ka*n alltså inte d[et.]
		you[just can't do th[at.]
57	chair	[ja
		[right

58 soc1 [me ja] förstår= ((till Anita))
[but I] do see= ((to Anita))
59 chair =vi k↓ommer runt till dig här så småningom [åså]
= we'll get r↓ound to you here soon [so]
60 Anita [mm]
61 chair me:n (.5) alltså för å >hjälpa mej< me de här me snäll (.)
bu:t (.5) well to >help me< out with this nice thing (.)
62 å nördar >alltså< (.) eh (.) när du ((till soc1)) säger snäll
'n nerds > right< (.) eh (.) when you ((to soc1)) say nice
63 eh (.5) hehhum vill ja associera till saker som (.9) e:h
eh (.5) hehhum I 'm associating to stuff like (.9) e:h
64 asså e.h (.5) nåt från femtitalet eller så man
well e:h (.5) something from the fifties or like people were
65 var sn*ä*ll och så [va] men e: (.8) eh (.) menar du eh
being n*i*ce like that [right] but e: (.8) eh (.) do ya mean eh
66 soc1 [°mm°]
67 chair (.)asså eh (.) öh den lagliga v*ä*rden eller va- >vad menar du<
(.)well eh (.) eh the lawful v*a*lues or wha- >what do ya mean<
68 med snällhet °å så där°° ((avtagande))
by niceness ° 'n stuff°°((trails off))
69 soc1 mm men men d*e*: (.2) e: [(.)
mm but but th*a*:(.2)t's:[(.)
70 ? [rrm ((clears throat))
71 soc1 >*al*lså de< *ut*trycker ju An*it*a s*jä*lv också n*u* här
>*w*ell it's< An*it*a her*sel*f expresses this too here n*o*w
72 i de här då [när] de-
in this one like [when] it-
73 fam [me-]
[bu-]
74 men *me*nar du män mänskor som vill Anita v↑*äl* eller sna-
but do ya *mea*n peo people who wish Anita w↑*ell* or ra-
75 (.) de bli ju lit [( )]
(.) that's a bit [( )]
76 psy [alls]å *ja* måste *ja* måste också bryta in
[wel]l *I* have to *I* have to interrupt too
77 för de *fi*nns ju d*o*m som (.2) ja tror att du har haft en
cause there are those who (.2) I believe you've hade a
78 b↑*ra* relati*o*n till (.) som du *in*te har beskrivit för m↑*ig* som
g↑*oo*d rel*at*ion with (.) who you *hav*e'nt described to m↑*ee* as
79 *tön*tar (.) som har varit sn*ä*lla (.) stöd (.5) *ut*omstående.
*geek*s (.) who have been n*i*ce (.) supp*o*rt (.5) *out*siders.
80 soc2 men de kanske e samma (de bli ju intressant) [( )
but maybe that's the same (that'd be interesting [( )
81 chair [ja: (men
[ye:(but
82 vi kommer *ru*nt så småningom men Greta ((soc1)) du...
we 'll come r*ou*nd there soon but Greta ((soc1)) you...

In brief outline, one of the social workers (soc1) introduces a concern with Anita's reluctance towards people who are trying to help her, people she allegedly views as 'nerds' (lines 5-11). It may be noted in passing that the social worker prefaces her presentation of the problem by offering a categorization of the issue as 'a bit special' (line 3), which provides for a hearing of the problem as a sensitive matter.[11] In line 11, she concludes the presentation with a rhetorical question ('I mean, where's that come from?'), upon which follows a clarification sequence in which the category 'nerd' is unpacked and which evolves into a humorous event in lines 18-20.

Possibly as a result of this side-tracking, the social worker refocuses the meeting on the issue she has introduced (line 24), now rephrasing the problem as something of direct relevance and interest for the upcoming assessment, by asking how it would be possible to give Anita proper support, should she one day feel the need for it. This time around, Anita responds to what she clearly hears as an accusation, by invoking a set of emotional states to account for her unfortunate statement that people who are trying to be good to her are nerds (lines 32 through 39). This account seems designed to waive the relevance of the social secterary's point, in that is suggests that Anita does not actually feel that way at all –she only said so because she was upset at some distant point in time.

Several features of Anita's account deserve further attention: for instance, the fact that she gives up her first oppositional attempt in line 28 suggests that she is not prepared to interrupt the social worker. Instead, she projects a new transition-relevant point in the social worker's turn in line 29, and then starts off by placing in terminal overlap a token of qualified agreement ('yeh but'), then by inserting a fast-paced self-selecting device 'but may I just say one thing' which allows her to continue with the account. The latter device can be seen to manage the 'backtracking' of Anita's turn; she is observably not attending to the previously adjacent turn, but rather seems to respond to the gist of the accusation presented much earlier (line 9).

Secondly, the self-selecting device also seems to reveal Anita's orientation to the formal speaking agenda which, in taking a turn at this point in the meeting, she can be seen to violate. Interestingly, her action seems to account for the violation itself: if the accusation is somehow misguided – which seems to be the upshot of Anita's account – then it is also irrelevant to the meeting and therefore requires immediate correction. This is further evident in line 37, where Anita rounds off the account suggesting that the concern raised by the social worker merits no further comment (although she does comment upon it in the next turn).

In other words, Anita specifically designs her action so as not to interrupt the social worker, orienting at the same time to the formal speaker agenda that she temporarily suspends. Whereas several of the participants seem to express some form of agreement with Anita's account (and nobody seems to object to her speaking), the social worker does not drop the issue. Instead, she reintroduces the issue of people being nice to Anita in lines 43-44, and begins to state the problem anew (line 46). However, she cuts off her turn and we see Anita taking another opportunity to account for her alleged behavior. This time she does so by employing a script formulation through wich she exploits a set of *if* → *then* relations (lines 47

[11] One of the way that this is accomplished (apart from the obvious observation that the issue is de facto prefaced) is through the laugter that embeds the final part of the social worker's turn in line 3 (cf. Osvaldsson, in press, for a discussion of the use of laughter in socially sensitive exchanges.)

through 51), that provide the 'basis for [her] accountability within a normative and moral order' (Edwards, 1995, p. 319-320).[12]

The question we need to address here is whether Anita's account, and more specifically its sequential placement in the midst of the social worker's turn, comprises a conversational event that we may warrantably describe as interruption. Certainly, the transcript clearly shows that the social worker abandons her turn and that immediately thereafter Anita begins to speak. However, research on turn-competitive incomings (French & Local, 1983, Schegloff, 2000) shows that competitive interjacent turns typically show various design features that will allow them to outlast an overlapping stretch of talk, such as increased amplitude, pitch, emphasis and shifts in voice quality. In the present case, the initial part of Anita's turn shows relatively few such features, which makes it rather hard to claim that she is in fact competing for the turn. Moreover, by the time Anita starts talking, the social worker has already cut off her turn, so there would seem to be little need for any turn-competetive measures on Anita's behalf, as there simply is no ongoing talk with which she would competitively overlap.[13] Finally, and perhaps most importantly, there is no indication in the ensuing interaction that an interruption has taken place. For this reason, while acknowledging that at a first glance line 47 may seem interruptive of the social worker's talk, we will maintain that this is not necessarily the case.

Skipping ahead in the transcript we may note that in line 58 the social worker aims to continue the disagreement, but drops out in mid-turn as chair starts talking to Anita, pointing out that she will get her turn at talk in due time (which she swiftly acknowledges in line 60). Hence, he is explicitly invoking and acting upon the official speaker rotation procedure, making it clear that it is not Anita's turn to speak. Having made this point, chair turns to the social worker, questioning the way she has formulated the problem. More specifically, chair questions the notion of 'niceness' as either too vague or possibly even insuficiently relevant for the institutional business at hand.

Exploiting Conversational Interruption

We have seen in the previous section how the social worker has presented the meeting with a problem, a problem that might be a bit tricky for all the involved organizations (both referring and admitting institutions) to deal with. After an exchange with Anita in which she questions the relevance of this problem for the meeting, chair invokes the speaker agenda to terminate this line of interaction. Instead, chair proceeds by problematizing the way that the initial issue has been formulated, making it impossible for the social worker to pursue her topic, at this point. Instead, she is facing the task of clarifying, and arguably accounting for, her phrasing of the problem proper.

And it is now that our target phenomenon occurs. Let us consider again the final part of the transcript:

69	soc1	mm men men d*e*: (.2) e: [(.)

[12] We may parenthetically note that the preface to Anita's turn ('but it works like this right') does not merely project an opposition towards the social worker's turn; it projects precisely this type of explanatory device.

[13] Though admittedly, there may well be visual cues that led the social worker to give up her turn.

		mm but but th*a*:(.2)t's:[(.)
70	?	[rrm ((clears throat))
71	soc1	>*al*lså de< *ut*trycker ju An*it*a *sjä*lv också n*u* här
		>*we*ll it's< An*it*a her*self* expresses this too here n*o*w
72		i de här då [när] de-
		in this one like [when] it-
73	fam	[me-]
		[bu-]
74		men *me*nar du män mänskor som vill Anita v↑*äl* eller sna-
		but do ya *mea*n peo people who wish Anita w↑*ell* or ra-
75		(.) de bli ju lit [( )]
		(.) that's a bit [( )]
76	psy	[alls]å *ja* måste *ja* måste också bryta in
		[wel]l *I* have to *I* have to interrupt too
77		för de *fi*nns ju d*o*m som (.2) ja tror att du har haft en
		cause there are those with wh*o*m (.2) I believe you've had a
78		b↑*ra* relat*io*n till (.) som du *in*te har beskrivit för m↑*ig* som
		g↑*oo*d rel*at*ion (.) who you *hav*e'nt described to m↑*e* as
79		*tön*tar (.) som har varit sn*ä*lla (.) st*ö*d (.5) *ut*omstående.
		*geek*s (.) who have been n*i*ce (.) supp*o*rt (.5) *out*siders.
80	soc2	men de kanske e samma (de bli ju intressant) [( )
		but maybe that's the same (that'd be interesting [( )
81	chair	[ja: (men
		[ye:(but
82		vi kommer *ru*nt så småningom men Greta ((soc1)) du...
		we 'll come r*ou*nd there soon but Greta ((soc1)) you...

Without going into details, we have noted that the interactional project that chair seems to pursue deals with the unpacking of a gloss ('nice') that the social worker used to formulate one of Anita's problems. The first line of transcript above shows that the social worker attends to this issue as a source of trouble, displaying several tokens of hesitation in the initial part of her reply (line 69). However, in line 71 she does respond to the chair's query, by pointing out (albeit not without hesitation) that those are Anita's own words, for which –by implication- she (soc1) can not be held accountable. In other words, the social worker's turn in lines 71-72 provides a way of relaying the burden of accountability back to Anita. However, the social worker's hesitant attempt to pinpoint exactly when or how Anita expressed her attitude towards 'niceness' fails, as the family counsellor places a potentially disaffiliative bid for the floor in the midst of the social worker's turn (line 73). And indeed we find the family counsellor disattending the social worker's trajectory of reasoning. Instead, she paraphrases chair's request for an elaboration of the social worker's (rather than Anita's) use of the gloss, by beginning to offer in line 74 two candidate hearings of the social worker's formulation. However, she fails to complete the second alternative, and moves on to begin assessing the first (and only) displayed hearing. In other words, the family councellor's turn can be heard as an attempt to reinstate chair's interactional project: to hold the social worker accountable for her earlier phrasing of Anita's problem. In this sense, it is clearly disruptive of the social worker's attempt to manage or waive this accountability by invoking Anita's own words.

The family counsellor's assessment of the social worker's problem formulation, or rather of its displayed interpretation, is in turn overlapped by the psychologist in line 76, in which she argues that Anita does in fact have some successful relations to people who have been nice, people whom Anita has not described as geeks. Thus, abandoning the issue of terminology, the psychologist addressess, and arguably challenges, the very problem initially raised by the social worker. In other words, just like the family councellor's turn seemed disruptive of the exchange between the social worker and chair, so does the psychologist now seem to disrupt the interaction between the family councellor and the social worker.

Let us take a closer look at how this disruption is accomplished. We have already noted that the psychologist starts off in overlap with what becomes the terminal part of the family councellor's turn. Because of this, it is impossible to tell whether the family councellor actually finishes her assessment or whether she drops out of the overlap abandoning the rest of her turn. Focusing instead on the initial part of the psychologist's turn, we may note that she starts off with the discourse marker 'alltså' (approx: 'well'), which in this position is hearable as a disagreement token, followed by a restart of what becomes a preface to her turn: 'I have to I have to interrupt too' (line 76). As Schegloff (1987) has shown, recycling of initial turn elements provides a way of dealing with simultaneous talk without having the central part of one's turn distorted by the overlap. In the present case, the recycling of the element 'I have to' in the psychologist's turn would seem to be a measure towards a projected stretch of overlapping talk. Such overlap could possibly involve the family councellor, should she opt to continue her turn. Alternatively, it could involve the social worker, to whom the family therapist's turn was allocated in line 74. In either case, the initial part of the psychologist's turn seems specifically designed to secure a turn space right there and then.

Apart (but not irrespective) from matters of actual and projected overlap, we need to consider in some detail the interruptive features of the psychologist's turn. Note that its preface explictly casts her actions as interruptive of the ongoing interaction: 'well I have to I have to interrupt too'. Moreover, in suggesting that she is (merely) joining the interrupting, the psychologist also construes the family councellor's turn as interruptive of the interaction between the social worker and chair. Seen as a way of claiming exclusive rights to the conversational floor, the psychologist's move is parasitic of the preceeding events, in that it displays a hearing of previous actions (in lines 76-77) as interruptive and at the same time uses that interpretation to legitimize a similar form of action within the normative order of the speaker agenda. It provides a way of creating a recognizably interruptive sequential environment. We will return to this matter shortly. Let us first point out that in contrast to the earlier event in lines 47-51, where Anita started to talk seemingly at the expense of the social worker's turn, yet where an analytical characterization of the event as interruption could not be sustained, the current exchange shows interruption to be a members' phenomenon, oriented to and invoked for local interactional purposes.

So what are the interactional upshots of the psychologist's actions, or more specifically, of designing her actions as warrantably interruptive of the ongoing exchange? We will suggest four distinct yet mutually interrelated interpretations. First, we have already seen that in the immediately preceeding turn (lines 73-75), the family therapist clearly addressed his query at the social secretary, allocating the next turn-slot for her reply. However, the design of the psychologist's turn beginning, and its sequential placement, is tailored so as not to allow this expected event to take place. Hence, the question whether the psychologist's turn is *de facto* an interruption of the family therapist's turn at talk is of little importance. Rather, its interruptive design (including its self-categorization as an interruption) provides for its illicit occupation of a pre-allocated turn space. If interruption is to be understood as a form of violation of participants' conversational rights and obligations (Bennett, 1981; Eglin, 2000; Murray, 1985), then the present case shows us that, although there is no way to conclude whether or not the family counsellor's right to complete her turn has been violated, at least the normatively expected order of speakers has been set aside. Indeed, a plausible hearing of the psychologist's 'well I have to interrupt too' is that it is designed to void the prior turn of its sequential implicativeness on the grounds that it was interruptively illicit to begin with. Again, it is not necessarily that turn itself that is subject to interruption, but rather the potential exchange it might have initiated.

Second, the design of the psychologist's turn allows her to artfully manage the importance and immediate urgency of her action. This is indicated not through the preface *per se*, but rather through the way that it is linked to the rest of the turn: 'well I have to I have to interrupt too *cause* there are those who...' (lines 76-77). A plausible hearing of the junction between the preface and the turn proper then, is that of a causal relation: interruption is necessary *because* what comes next is crucial to the currently ongoing interaction. As such, it can not be delayed. Hence, the causal link also provides an account for the interruption. Indeed, it renders interruption a necessary form of delivery, considering the implied acute consequences of what she has to say for the business of the meeting.

Third, and in relation to the above, we have noted that rather than continuing the topic and argumentative trajectory of the previous turn, the psychologist refocuses on the nature of Anita's acclaimed problem. This shift of topic clearly allows her to discontinue the negotiation of what type of behavior may be glossed as 'being nice to Anita'. Another way of accomplishing this shift of trajectory has to do with the allocation of the psychologist's turn: Although the exchange she interrupts takes place between the social worker and the family councellor, it is to Anita that she directs her talk. If we accept that the psychologist's turn is tailored to be received as crucial to the ongoing interaction, we now have to conclude that as far as her actions are concerned, the important business has to do with Anita, rather than with the family councellor's (and presumably the chair's) splitting hairs on terminology. Although somewhat speculative, this interpretation seems to be supported by the fact that the psychologist actually recycles the 'problematic' term ('who have been nice', line 79) in a completely unproblematizing manner. Thus she is demonstrably settling for a commonsense understanding of the notion of 'niceness'. In effect, the psychologist's turn may be seen as designed to be received as acutely important to the meeting –hence the interruption. Yet, having secured her turn space, she thoroughly disattends the previously ongoing exchange. In this sense, we would argue that the psychologist's actions are disaffiliative with the preceeding interaction, and seem to disqualify its relevance to the institutional concerns of the meeting.

Fourth, we have already noted that the speaker agenda is a relevant concern for the participant's, whose actions reveal an orientation to the normative speaking order at the meeting (e.g., Anita's turn beginning at lines 31-32). We now wish to suggest that in treating the previous turn (lines 73-75) as interruptive of the exchange between chair and the social worker, the psychologist constructs and acts upon an interruptive sequential environment. An evident feature of such an environment is that it by necessity brackets the official speaker agenda. Thus, the psychologist's turn preface in line 76 is designed not only to locally manage its occupation of a pre-allocated turn space -it also sems to deal with the fact that the psychologist is not adhering to the speaking order. On this view, the psychologist can not be accountable for violating the agenda, since it has already been suspended by the family councellor's interruption in the preceeding turn. In other words, the turn preface constructs the psychologist's action as a warrantable event within the local order of events. A final indication that the speaker agenda needs to be attended to and locally managed is to be found in the very last turn of the transcript, in which chair invokes the rotating speaking order as a way of disrupting the second social worker's turn (line 80) and by the same token, reinstalling the formal speaker agenda.

CONCLUDING DISCUSSION

In this paper, we have addressed at some length various issues of overlapping speech and interruption, focusing in particular on how these conversational phenomena are dealt with by participants in the turn-by-turn unfolding of spoken interaction. Using empirical extracts from two rather disparate interactional settings, we have aimed to highlight three analytical points.

First, reviewing some influential conversation analytic studies of overlapping talk, which overwhelmingly deal with dyadic and -to a lesser extent- triadic interaction, we have attempted to indicate some ways that this strand of inquiry may be extended to include interactional exchanges involving a greater number of potential conversational parties, as well as more complex participation frameworks, that is, constellations of speakers-recipients. Here, we have shown that participation frameworks in multiparty exchanges may result in more complex and hitherto unexplored patterns of overlapping talk.

Second, our analysis of overlapping talk has led us to attend to a few instances of conversational interruption. Here, aligning with an ethnomethologically informed, participant-oriented perspective on talk-in-interaction, we have argued that an instance of conversational interruption – as relevant and locally consequential to the particpants' conduct – can not be analytically established purely on the basis of structural features of conversation. Rather, for analysts of conversation interruption can only be identified through participants' own assessment of a prior or concurrent action, as recognizably disruptive of the current flow of interaction (Bennett, 1981; Bilmes, 1997; Eglin, 2000; Murray, 1995). Accordingly, our analyses demonstrate different ways through which participants at talk go about 'being interrupted'. Notably, we have shown that establishing an incoming turn as interruptive of currently ongoing talk is not necessarily a task for the current speaker, but may well be carried out by parties engaged in 'doing recipiency'. Indeed, we would argue that an analytical procedure that treats interruption as merely a matter for speakers is incompatible with the dialogic conceptualization of language and social interaction which informs all analytical approaches derived from ethnomethodology. On this view, it is not just the speaker who is interrupted – nor, for that matter, the recipent. What is interrupted is inescapably the *interaction* between two or more parties at talk. This highlights the importance of studying the organization of overlapping talk and interruption in multiparty setting, since analyses of dyadic interaction – where both parties engaging in overlap clearly are speakers – naturally overlook this point.

Third, and following from the above, is that interruptions are interactively handled. Not only are they jointly accomplished; as interactional events they may be exploited for local interactional purposes. As Eglin (2000) points out: 'a recognized interruption can be defeated, refused, reduced to an attempt, forestalled, *turned to advantage*, *incorporated*, let pass, ignored' (p. 64, emphasis added). Accordingly, our analysis of Extract 2 has shown how a participant to a formal meeting may pursue a range of local interactional goals by constructing a previous turn as interruptive of the ongoing interaction. In other words, interruption is not necessarily only a method for gaining an immediate turn at talk –it is a form of action with its own local indexical features and reflexive upshots. As such, it forms a members' method for the construction and interpretation of coordinated trajectories of locally accountable actions.

Finally, taking into account earlier studies of interruption, we would not wish to deny its relevance to our understanding of power and dominance (Zimmerman & West + fler), nor to to social proximity and engagement (Tannen, 1982). However, we would maintain that a thorough examination of the local empirical bases conversational interruption is analytically prior to such considerations. It is only then we may find out *how* power asymmetries, gender differences, social and cultural proximity or distance, or any other aspect of social life prone to theorizing, are invoked, exploited and created by members-in-society acting in the realm of the natural attitude.

REFERENCES

Bargiela-Chiappini, F. & Harris, S. J. (1996). Interruptive strategies in British and Italian management meetings. *Text, 16*, 269-297.

Bennett, A. (1981). Interruption and the interpretation of conversation. *Discourse Processes, 4*, 171-188.

Berger, C. R. (1994). Power, dominance and social interaction. In M. L. Knapp & R. Miller (Eds.), *Handbook of interpersonal communication.* (pp. 450-507). Thousand Oaks, CA: Sage.

Bilmes, J. (1997). Being interrupted. *Language in Society, 26*, 507-531.

Cromdal, J. (2001a). Overlap in bilingual play: Some implications of code-switching for overlap resolution. *Research on Language and Social Interaction, 34*, 421-451.

Cromdal, J. (2001b). 'Can I be with?': Negotiating play entry in a bilingual school. *Journal of Pragmatics, 33*, 515-543.

Cromdal, J. (2004). Building bilingual oppositions: code-switching in children's disputes. *Language in Society, 33*, 33-58.

Cromdal, J. & Aronsson, K. (2000). Footing in bilingual play. *Journal of Sociolinguistics, 4*, 435-457.

Drummond, K. (1989). A backward glance at interruptions. *Western Journal of Communication, 53*, 150-166.

Edwards, D. (1995). Two to Tango: Script formulations, Dispositions, and Rhetorical Symmetry in relationship troubles talk. *Research on Language and Social Interaction, 28*, 319-350.

Eglin, P. (2000). Interruptions. *Ethnographic Studies, 5*, 56-67.

Ferguson, N. (1977). Simultaneous speech, interruptions and dominance. *British Journal of Clinical Psychology, 16*, 295-302.

French, P. & Local, J. (1983). Turn competitive incomings. *Journal of Pragmatics*, *7*, 17-38.

Garfinkel, H. (1967). *Studies in ethnomethodology*. Englewood Cliffs, NJ: Prentice-Hall.

Goodwin, M. H. (1990). *He-said-she-said*. Bloomington: Indiana University Press.

Goodwin, M. H. (1998). Games of stance. In: S. Hoyle & C. T. Adger (eds.), *Kid's talk: Strategic language use in later childhood*. Oxford: Oxford University Press.

Gumperz, J. J. (1982). *Discourse strategies*. Cambridge: University Press.

Heritage, J. (1984). *Garfinkel and ethnomethodology*. Cambridge: Polity Press.

Hopper, R. (1992). *Telephone conversation*. Bloomington: Indiana University Press.

Huls, E. (2000). Power in Turkish migrant families. *Discourse & Society*, *11*, 345-372.

Hutchby, I. (1993). Confrontation talk: Aspects of 'interruption' in argument sequences on talk radio. *Text*, *12*, 343-371.

Jefferson, G. (1983). Notes on some orderliness of overlap onset. In V. D'Urso & P. Leonardi (Eds.), *Discourse analysis and natural rhetoric*. (pp. 11-38). Padua: Cleup Editors.

Jefferson, G. (1986). Notes on 'latency' in overlap onset. *Human Studies*, *9*, 153-183.

Lerner, G. H. (1996). On the 'semi-permeable' character of grammatical units in conversation: conditional entry into the turn space of another speaker. In: E. Ochs, E., A.

Schegloff & S. A. Thompson (Eds.), *Interaction and grammar*. (pp. 238-276). Cambridge: University Press.

Lerner, G. H. (1989). Notes on overlap management in conversation: The case of delayed completion. *Western Journal of Speech Communication*, *53*, 167-177.

Murray, S. (1985). Toward a model of members' methods for recognizing interruptions. *Language in Society*, *14*, 31-40.

Osvaldsson, K. (2002). *Talking Trouble. Institutionality and Identity in a Youth Detention Home*. Linköping: Linköping studies in arts and science, 263. (Diss.).

Osvaldsson, K. (2004/in press). On laughter and disagreement in multiparty assessment talk. (To appear in TEXT.)

Palmer, M. (1989). Controlling conversations: Turns, topics and interpersonal control. *Communication Monographs*, *56*, 1-18.

Rogers, W. T., & Jones, S. E. (1980). Effects of dominance tendencies on floor holding and interruption behavior in dyadic interaction. *Human Communication Research*, *1*, 113-122.

Sacks, H. (1992). *Lectures on conversation, Vol. 1*. Gail Jefferson (Ed.), Oxford: Blackwell.

Sacks, H., Schegloff, E. A., & Jefferson, G. (1974). A simplest systematics for the organization of turn taking for conversation. *Language*, *50*, 696-735.

Schegloff, E. A. (2000). Overlapping talk and the organization of turn-taking for conversation. *Language in Society*, *29*, 1-63.

Schegloff, E. A. (1987). Recycled turn beginnings: A precise repair mechanism in conversation's turn-taking organisation. In G. Button & J. R. E. Lee (Eds.), *Talk and social organisation*. (pp. 70-85). Clevedon: Multilingual Matters.

Schegloff, E. A. (1972). Sequencing in conversational openings. In J. J. Gumperz and D. Hymes (Eds.), *Directions in sociolinguistics*. (pp. 346-380). N.Y.: Holt, Reinhart & Winston.

Schegloff, E. A., Jefferson, G. & Sacks, H. (1977). The preference for self-correction in the organization of repair in conversation. *Language*, *53*, 361-382.

Tannen, D. (1982). *Conversational Style*.

West, C. (1979). 'Against our will': Male interruptions of females in cross-sex conversations. Annals of the New York Academy of Sciences, 327, 81-97.

West, C. & Zimmerman, D. H. (1983). Small insults: A study of interruptions in cross-sex conversations between unacquainted persons. In B. Thorne, C. Kramarae & N. Henley (Eds.), *Language, gender and society*. (pp. 102-117). Rowley, Mass: Newbury House.

In: Advances in Communications and Media Research, Volume 4 ISBN 1-60021-189-5
Editor: William H. Miller, pp. 129-150

Chapter 7

The Media Truth Commission: Brokering a Healing History

Charmaine McEachern
Adelaide University, South Australia

Introduction

As an extended shamanising media event, the intention of which was to promote reconciliation and healing as a primary process of nation building, the core of the media TRC can be seen as pedagogic, providing a new frame and purpose for the nation and its institutions. As in all public events (Handelman 1990) this desired interpretation was contained in the very sequencing and ordering of the event, in the form of the three committees which both generated a moral process and enacted a model of healing for individual and nation. This moral process was constructed within at times contested discourses of truth and reconciliation, with imagining and promoting a necessary unity underpinning the whole. The outcome of the whole process constructed as a media event was a narrative of the nation which, given its preoccupation with the past can also be understood as an alternative history, a new, inclusive representation of the past which could serve the interests of nation building.

At one level, such a construction in itself constitutes the nation as an "historical subject with a 'national past'" (Malkki 1995: 257), so that the media event vacillated between the interpellation of nation as already existent and the language of nation building, the nation yet to emerge, to be "born" (Mona 1998: 38). Central here, and in the face of objections discussed in the previous chapter, history took the form of a representation of the apartheid past, rendered as a history of the *whole* nation, not just the part, as the gaps, the denials and the silences of the past were filled in. In this form history, as a representation of the past, is constituted out of memory, the sedimentation of past experience, which is at the same time responsive to present needs (Isla 1998). The moral process thus refigured the *meaning* of the past in very particular ways, which as Handelman argues, is exactly what history entails (1990: 225). In turn, this history owed a great deal to the TRC itself, its forms, aims, the kind of evidence it used (confession and testimony) and its own constraints, not the least of which was its public media form.

The TRC and thus the media TRC, is transformative in the sense that shamanising media events are. This, as Bozzoli argues, was its goal; to transform wounded individuals into whole people and to transform the divided nation into a new unity. The "proposal for radical change" (Dayan and Katz 1992: 153), which is the intention of shamanising media events, here is pedagogic, teaching the people (including institutions) knowledge which has the specific aim of changing them and the basis of nation itself. Here, the new paradigm for society, its proposal for a new moral order, is founded on knowing the past in order to turn away from it. This new self-knowledge then becomes the basis for society and nation.[1] At the transformative core of the event, the refiguring of history as the meaning of the past, was to make possible a new beginning, overturning old paradigms and ideologies.[2] As a consequence, history itself was encompassed by morality and knowledge was conceived as truth.

Dayan and Katz argue that shamanising media events force new ways of looking and thinking from within their own performance. The extended nature of the media TRC means that this performance was further extended, in time, field and complexity, offering possibilities for new ways of thinking which were not necessarily totally contained by the public event itself. In his comprehensive exploration of public events, Handelman (1990) makes a distinction between the intention and design of public events and their actual performance. For him, most of the ritual purpose and purchase of such events is in the former, so he devotes little attention to the latter. In media events, where media have to be factored into the design of the public event or process, I suggest that the analyst has no such option, for much of the purpose of the event has to be envisaged in terms of performance itself. This is particularly significant where the object of contemplation and construction is the nation. How the nation is imagined and through what forms of solidarity and identity are crucially connected to the dominant media forms that carry imaginings of collectivity and solidarity.

In this regard, the media TRC becomes transformative at other levels as well as that of the event itself. Media introduces a kind of transformative redundancy, or a chain of transformations which ultimately affects the TRC as a public event, particularly its ability to imagine the national collective; 'the people'. The media TRC is, as Becker argues is true of all mediated forms of public event (1995: 634), further transforming, in its performance. As Chapter 2 established, the public event of the TRC is transformed into media event through media genres, techniques and conventions. Rather than just representing the TRC which happens elsewhere, media are central in performing it as a public event (cf Becker 1995: 640), shifting it from the purely ritual sphere to the media one, and in the process, transforming it into media performance of public ritual. In this public performance of event we find MacAloon's "nesting of performance genres" (cited in Handelman 1990: 59), a process in which journalistic conventions were particularly dominant, given the political

---

1 Turning away from the past does not of course equate with forgetting it, which is why the CSVR have published studies on the question of for how long South Africa will need to listen to the voices of those harmed under apartheid (see for example, Hamber 1997b)

2 Here South Africa becomes a fascinating example of White's performance of the collectivity through "public historical discourse":

> Asking how collectivity is not only imagined, but also how it is felt and performed in public historical discourse poses useful questions about the social and cultural architecture of national memory - seen as an achievement in the making, rather than a finished product (1997: 5).

nature of the extended media event. The reflexivity and transformations involved in the TRC are importantly the outcome of the media's *performance of* the TRC, which means that it is this performance, the media text of the TRC, which requires deconstruction.

Such multiple levels of transformation must inevitably be dangerous and uncertain (cf Handelman 1990: 31) in terms of outcome, and this is especially so in the media TRC. The media performance of the TRC's pedagogic imagining of nation introduces interpretation, diversity and re-presentation (Morse 1998: 54), all of which have the potential to disrupt and alter the official imagining of the public event. The selections, interpretations and the transformations which media conventions and generic forms introduce are all important in this regard. At the same time, as this chapter demonstrates, the re-presentations themselves, as transformations that are *also* informed by the pedagogic intentions of the TRC as a public event (see Chapter 2), are central in narrating the nation, elaborating and enhancing some of the techniques which were introduced into the ritual to produce its shamanising effects. In particular the chapter argues that media practices and conventions were central in the *moral* inflection of the new historical narrative emerging in the TRC. At its heart, the media performance produced the uncanny, the destabilisation of the familiar, which was an important condition for the construction of a new imagining of the nation as 'the people', a unity. First, the uncanny makes possible a re-evaluation of official accounts of the past out of which can emerge a new, shared, history, as an inclusive narrative of 'the people'. Second, as an effect of the media's performance of the TRC, it dissolves those boundaries which Thornton saw preventing any emergence of nation, stressing sameness rather than difference. The chapter demonstrates that the uncanny, as an outcome of the media TRC, was central to the generation of the TRC's new paradigm for society, the core of the shamanising event. It locates the construction of the uncanny in an exploration of the media TRC as an outcome of media performance and interpretation.

South Africa's TRC as Broadcast History: Revelation and Healing

To understand what the TRC was doing and why particular dimensions emerged and were elaborated in its media performance, it is useful to start with a few simple propositions about the apartheid state in its most self protective, totalising moment. What is important about the South African state in the sixties, seventies and eighties, which is most relevant to the TRC, is not so much the racialised basis of apartheid, though this underpins all. Rather it is the fact that the state could organise and institute apparatuses that tortured and killed in the certain knowledge that this would be kept concealed. Doctors, judges, lawyers - the media - would in the main, endorse these acts, as part of the routinised practice of the combined apparatuses of the state. Often clearly based on bad faith (for example, the explanations in the Steve Biko case were clearly implausible), this immunity of the state from public knowledge of its actual provenance was fundamental to its power and legitimation (Nursey-Bray 1994: 9).[3] Here lies the fundamental paradigm shift at the moral and political core of the TRC as a

3 Race is a factor here of course. Nursey-Bray points out that the "coercive authority" of the state was almost inevitable given its problems of legitimacy among the non-white population. Yet, at the same time, the state required concealment of this coercion to maintain white consent.

media event; in *its* terms such state power was illegitimate and unlawful and previously immune officers of the state were to be held accountable, at the very least at the level of explaining their part in the apartheid system of domination.[4] As such it clearly signalled a break with the past, even as the past was its focus. This break both authorised claims about the 'newness' of the nation and set in motion the re-interpretation of history itself as a narrative of the past. At the same time the realities of the apartheid state were exactly what made the past unknown, unknowable in any normal, document based process of gathering history. Rather, history had to be built on memory and experience, a process which had the additional virtue in a postcolonial situation, of providing space for those usually left out of official histories. Inclusivity in history symbolised the inclusivity that was to characterise the 'new' nation.

So what was the narrative of the past which emerged in the TRC as a media event? What were the narrative elements that constituted this new story? And, how were its narrative elements constructed, inflected and shaped through the genres and conventions of media themselves? I have argued that journalism and journalists were dominant in the media TRC. Routinely, it was news and current affairs forms and conventions that dominated and stage-managed the performance of the media event, articulating these with other genres to construct the whole. This dominance is not unusual. For example, in his analysis of the Gulf War as a media war, Kellner (1995) argues that not enough attention is paid in the analysis of media events to the effects of routine practices of news reporting. In the media's performance of the TRC, the concept of news values[5] was particularly efficacious. In media terms, the TRC was what journalists see as 'good news' on many counts. It was a local story, relevant to the entire population, it was a story of extraordinary things happening to ordinary people and it also had stories of great men and women, some of whom had been elevated to the highest positions in the land. Most particularly, it was the story of conflict; conflicts of all kinds. It was, as Krog says, "the story of the century" (1998: 46) for all of these reasons.[6] It had drama,[7] human emotion and travail. These are all primary news values in contemporary news practice and the narrative emerging in the media TRC proceedings was elaborated and framed through all of them. Interestingly too, the media event shared a common form with journalism, the centrality of narrative itself. Many theorists have noted how journalists in news media actually produce their 'news stories' through the narrativisation of events that are construed as 'news events'.[8] In the media TRC journalists had a wealth of narrative material with which to weave their own story of the event.

---

4 This was clearly part of the objections from the South African Police Force (SAPS), voiced in submissions to a conference on setting up a truth commission, initiated by Justice in Transition (see Boraine and Levy 1995: Appendix 1). It was exactly the reliance upon notions of the normalised legitimacy and legality of the apartheid state operations to which they appealed when they opposed a truth commission, arguing instead for dealing with the past through the law courts. This was the official line of the NNP throughout also.

5 See Gruneau and Hackett 1995; Ward 1995; Hartley 1987; Allan 1999.

6 One should point out that Krog makes this comment in the context of talking about foreign journalists' reactions to the TRC.

7 The relevance of drama to news footage, which is usually taken for granted by news personnel, was made apparent during televised reporting of the first Cape Town hearings when the newsworthiness of the hearings was compared with events outside, with a link in voice over that said, "The drama was not confined to the hall" and revealed with footage, that the car carrying Deputy Chairman Alex Boraine had collided with the back of that of Chairman Tutu on the way to the TRC that morning.

8 See Bird and Dardenne 1988; Jacobs 1996; Ward 1995.

The intention to interrogate the past and the notion of rupture provided the media focus right from the start of the TRC. These principles established a kind of encompassing frame for the media's narration of the new story of the nation, at the same time setting the narrative in motion. What emerged was a dominant narrative theme that the TRC *was* history, itself symbolising the break with the past that the state claimed for the new nation. The first week of hearings in Eastern Cape was formative. Here the coverage set in train the moral narrative of the TRC as a media event, literally as a beginning after rupture (Werbner 1998), the start of a great event in South Africa's history and very much in keeping with the shamanising dimensions of this media event. We can clearly see the theme of the TRC as itself history in the making in the page one coverage of these hearings in the *Mail and Guardian*, which made very strong statements that "History is being made" and that the victims of apartheid were carving out a space *in history* for themselves. The first television *Truth Commission Special Report* also took up this narrative theme, and introduced it as the perspective of those testifying, at one stage including a comment selected from an interview with Karl Weber, victim of the Pan African Congress's (PAC's) Highgate bombings, that the TRC is "the beginning of something". Part of the media event's reflexive terrain for viewers and readers (Becker 1995: 642) was thus the historic nature of the TRC itself. The media TRC wasn't just broadcast history (Dayan and Katz 1992: Chapter 1), it also reflexively interrogated itself *as history*, a comment on the very nature of this process which was producing the new official history.[9]

In such a construction the TRC itself symbolises and enacts South Africa's postcolonial moment. Implicit in the representation is the notion that testimony, the dominant form of speaking at the human rights hearings being reported, is resistance to oppression, particularly in its form of silencing the oppressed. Supriya's (1996) comments on the theory of testimony developed by Felman and Laub in their work on Holocaust witnessing, are suggestive here. She notes that testimony must be seen as:

> ...a rhetorical genre, through which the witness even while appearing to be inextricably positioned in relation to his/her historical experience of oppression, is paradoxically liberated from the coercive force of that very same history...this is so because the very discursive act of the testimony and the testimony itself resist the oppression that once and seemingly eternally condemned the witness to silence (1996: 99).

In the media's constructions, what was stressed was that on their television screens and through their radios audiences were seeing and hearing people who were speaking into the silences and gaps of history, against the misrepresentations of the past, pitting memory and experience against powerful, state controlled knowledge which was really concealment. Witnessing and testimony enabled the historic impact of the TRC and, in the sense of the media reports, history is constituted as the break *with* apartheid history and that concealment. That which was erased and silenced in the nation's past is now publicly voiced to provide identity for the new nation.

All of these narrative themes around witnessing can be found in another early example from print news. Under the headline, "Heart-rending", the *Sowetan's* front page coverage

---

9 This reflexive feature is not confined to shamanising media events. Part of the meta-narrative of Princess Diana's funeral, undoubtedly a media event in Dayan and Katz's coronation category (1992: Chapter 2), also interrogated the history-making aspects of the ceremony and its media coverage.

(16.4.96) of the first sitting of the TRC in East London begins with the testimony of Mrs Nomhle Mohapi, widow of Black Consciousness Movement activist, Mpetla Mohapi, who was tortured and died in custody, moves to the case of Mr Karl Weber who was maimed in the PAC 1991 Highgate massacre and then to the widows of the Pebco Three, which is the focus of the rest of the piece. Here operate other news values and strategies, those of conflict and binary opposition. News stories routinely render events and the people involved in terms of antagonists with two sides pitted against each other in some way. In this account, state violence is juxtaposed with liberation struggle violence, though the former dominates. The photograph at the centre seems to encompass the whole, brokering a preferred, encompassing meaning through its choice and positioning (Hall 1980; Seiter 1999: 20- 21), which draws together the narrative strands of witnessing. It is of Archbishop Tutu greeting Nohle Mohapi who is sitting in the row of people to give testimony at the front of the hall before the historic opening session. This photograph is emblematic of the whole and a visual version of the narrative as it is activated by the notion of witnessing (and listening). Here, ordinary people are for once explaining to the elites what it was like to live under a particular regime. In fact they are invited to do so and cordially welcomed, by those elites.[10] The photograph demonstrates that this is something different from (and more than) a legal process; this is not a law court.[11] It also captures the postcolonial in the TRC process. Here, that which had been private and marginalised under apartheid has been moved into the public sphere, and not just heard but also valued. Bozzoli makes this point about the hearings themselves, using Habermas to assert that the hearings can be understood as a radical version of the making public of the "proletarian realm" which is always "kept hidden, secluded, unheard" (1998: 168). It underpins the central, restorative justice (Girard 1972: 21) dimension of the hearings, but the *fact* of the media coverage further affirms it and the photograph at the centre in a sense makes the *process* of making visible, also visible.

If witnessing is what sets the media narrative of the TRC in motion, as history, then revelation provided a protracted middle which, day by day built up the content of the new history. Witnessing and revealing were interconnected in the coverage, producing what was represented in most accounts as 'truth'. Here the media were faithfully rendering a central tenet of the TRC as public event. Antjie Krog quotes Tutu providing the raison d'etre for the TRC by saying, "If you don't know the past, you will never understand today's politics" (1998: 171; cf Perelli 1994). What I have called the protracted middle of the media TRC narrative thus consisted primarily of the information and reinterpretation elicited in the testimonies of first victims and then perpetrators. The dominant narrative theme was revelation.[12] The media TRC was represented through revelations of all kinds; bizarre and horrific details of the torture of activists or of the abuse of suspected spies in ANC camps, murder and rape, the killing of detainees in gaols, cross-border raids and the destabilisation of

---

[10] Television exerpts of the hearings and radio coverage in particular were even better at establishing this sense, for here were recorded, in vision and sound, the courtesy and the respect accorded to testifying victims. Interestingly, there was more acrimony and an often sharper tone in the questioning of perpetrators in the amnesty hearings.

[11] Of course it was the official but not legal, in the sense of a law court, setting of the TRC which drew criticism from those opponents who noted the narrative form of the testimonies by arguing that they were "untested evidence" and therefore unsound (see for example Krog 1998: 215).

[12] This tenor of revelation clearly at times offended people who had been the victims of the apartheid state, who claimed that there was nothing new to them in the hearings (see Krog 1998; "TRC a whitewash of sins, says author", *Weekend Argus* 5.6.96; 9). I also heard this position voiced in discussions about the TRC, by some who had been activists in the 1980s in the District Six Museum, and in informal talks with township tourism guides.

countries adjoining South Africa, death farms, death squads and other state funded special units in every branch of the armed forces, planted evidence on corpses, propagandist blaming the ANC or the UDF for terrorist acts sanctioned by apartheid officials and the existence of 'third force' operations to destablise the country in the period before the 1994 elections. Many suspected deaths were also confirmed (where previously authorities had insisted the victim had left South Africa, maybe to join the ANC). Then there were the revelations of what had actually happened to the disappeared, or the farm or patch of veldt where some of their remains could be found. In all of this, media rendered revelation in their own terms, in terms of their own cultural performance and performance genres. Particularly influential here was the media's use of grand cultural themes in their narrative.

One such theme was the quest, a powerful media narrative theme that inhabits many media genres. In this media event, the TRC was turned into a narrative as a quest for truth to provide healing, which in a postcolonial context was enormously legitimating of the TRC, the state and the nation as well as the new narrative of the past. In the quest, what was important was not just that victims *provided* information, through breaking their silence of oppression in testimony, but they also *demanded* it by asking exactly what happened, when, where, and who did it? Here, witnessing remained oppositional, now challenging apartheid's legitimacy by testifying to the power of apartheid's concealment. Often people simply did not know what had happened to loved ones who disappeared, and that memory and the absence of closure was haunting. In shifting into this register of asking for, rather than offering revelation, people were granting a kind of legitimacy to the TRC and publicly marking the transition to a new state and a new time, where they could not only speak about the past but also find out its truths. At a practical level, this aspect of the testimonies offered a project and a narrative frame for journalists, since the kinds of questions people asked were of course exactly those to which journalists can relate. They form the basis of journalistic story-telling; the how, why, when and where of news events (Ward 1995; Allan 1999). A vivid example here is from the *Sowetan* (16.4.96: 11), under the headline, "Missing...without a trace". The article suggests that the TRC offers hope to families who lost members in the apartheid years and publishes photographs of some of these people (all male), accompanied by the stories of their relatives. The hope is that the TRC will be able to find out what happened to them.

The very first *Truth Commission Special Report* on SABC 2, presented by Max du Preez provides a good extended example of the role of revelation in the media's narrativisation of the TRC's events and just how media performance elaborated this in terms of quest. It took the case of the widows of the three Eastern Cape activists known as the Pebco Three which had had a high profile during the week's news. Du Preez linked the separate testimonies by saying that the widows, "would only be able to make peace with their lives once they have found out how their husbands were murdered". The testimonies provided to the TRC (and heard on radio) were long and detailed, covering the entire story of the disappearance, efforts to find out what had happened and subsequent details of how they found out their husbands were dead. At times they were hesitant, interrupted by weeping, and progressed in a non-linear way which occasionally blurred the sequence of events. To frame the program and establish the context for the television coverage of the TRC, one specific element out of this wealth of detail was selected. That was the image of one of the widows saying that she never buried her husband and that she wanted an investigation into his death (how, why and where) and another saying she too "would like to know the whole truth". Through du Preez's piece to camera and the edited segments from the testimonies, the notion of revelation provided the interpretation of the hearings, conceptualising revelation as 'the truth' and linked to healing. Thus this early television investigation of the TRC was itself initiated and framed through the TRC philosophy that truth (finding out what happened) produces reconciliation and healing. This philosophy was at times to become the centre-piece of the media performance and the means by which the media shifted themselves into the role of major actors *within* the TRC, something which emerged in this same program.

The rest of *this* particular show aligned *itself* with the TRC's quest for truth by inflecting revelation through news conventions and strategies. The program made the revelation that one of the killers was Joe Mamasela from the infamous Vlakplaas unit. He then confessed before the cameras and explained what happened in two of the most famous 'unsolved' cases from the nineteen eighties, the Pebco Three and the Cradock Four.[13] In turn, this television confession made news in *other* media, like radio and the press, the next day. Although it simply confirmed the suspicions of local people (Ergas 1994: 161-173) and much of the evidence had already been collected by journalists by the end of 1989, though they were unable to publish it (*Between the Lines* 19.5.96), this was news, sensational 'breaking' news, headline making news of the kind valued highly by media agencies. And it was news *created* by media agents, not simply reported. There are various features of media narrativisation which can perhaps explain this shift and its contribution to the media's telling of the nation's story.

In many ways, especially given its protracted nature as a public and media event, the TRC was not integrated and seamless, manifesting many points of disruption, discontinuity and fragmentation which at times obscured processes of cause and effect. Human rights violations hearings raised cases which were not answered or which could have no easy answer. At the same time, amnesty testimonies detailed cases which had not been raised as demands for truth. The duration of the TRC as a public event (Morse 1998) often separated out the various stages of the narrative strand of revelation into its initiating demands for 'truth' in the human rights violations hearings, then much later, the provision of the information by the perpetrators in the amnesty hearings. While the radio broadcast of the whole process maintained this duration, the media narrativisation of the TRC in other news forms often did not.[14] At the very least, images and footage from the human rights violations hearings were used to frame later amnesty hearings, so that these hearings were in part understood in terms of the contrast with the human rights hearings.[15] The strategy also shortened duration (though maintaining sequence) in the interests of providing narrative structure and closure.[16] In other cases, like the *Truth Commission Special Report*, investigative journalism emerged as a dominant form of cultural brokerage to do the same thing. The deferring of closure within the TRC process is problematic within investigative journalism, which was therefore able to use programs like this one, and the many

[13] Mamasela, who was originally an ANC member, recruited by the security forces for their special operations, actually refused to apply for amnesty and testify before the TRC, which significantly made the television forum *the* major public arena for his confessions and explanations.

[14] This is also a dimension of Peters' notion of the 'totalising' of media representations compared with individuals' 'fragments' of society and events (1997: 82).

[15] For example "'Kleenex' truth inquiry starts big clean-up", *Mail and Guardian* 16.8.96.

[16] Altheide and Snow (1992) point out that this is something that *all* news services do, drop file footage from past news event, here the original hearings, a wide range of other media narratives and contexts. Of course, there is then the possibility of producing different meaning as the original is anchored differently. Such is the nature of

documentaries and special press reports,[17] to construct and enact a narrative closure, at the same time shifting media *into* the TRC event and process in a different way.

Mamasela's confession set up a televised model for the desired truth of the TRC; a model of the world that anthropologist Clifford Geertz (1973) argues, tells people how things ideally *should* be. Subsequent television and press reports also 'solved' other cases, like the murder of Griffiths Mxenge[18] and the assassination of SWAPO leader, Anton Lubowski, seeking to clear him of the charge of working for the apartheid government. This was a response to a request by Lubowski's mother to the TRC to clear her son of the regime's insinuation that he had been a traitor to his political cause. In both these cases and many more, *media* aimed to use revelation to fulfil the requests of the victims to the TRC, aligning themselves with the process.

Documentaries, like the SABC 3 *If Truth Be Told*, also enacted the same role. A different kind of revelation was also performed, highly reflexively, when in current affairs style the *Truth Commission Special Report* deliberately brought together murderer and victim, Vlakplass member Paul van Vuuren and a child whose parents he killed (Thloloe 1998). Here media were actually staging and performing the dialogue between the victims and the perpetrators which many victims had asked for (van der Merwe 1998: 3).[19] Further, they were also drawing on and providing a different model for reconciliation itself, a different version of the reconciliation process (Hamber et al 1998: 2), which located victims *in* the amnesty process where many survivors felt they had actually been left out, sometimes forced to rely on media for accounts of cases in which they were directly involved and not having to forgive for the process to occur.[20]

In the media TRC, witnessing, revelation and confession articulate to produce truth. Truth then takes the form of a narrative, a morally inflected shared history, one which is owned and told by 'ordinary' people from all sides (Krog 1998: 51) rather than leaders, the privileged or the already famous. Through revelation and the re-naming of their experiences, the people take responsibility for history, a new postcolonial history which emerges as old, official history is first challenged, replaced by and then reconstituted through, witness, testimony and confession. Here the force of history turns on experience and memory, in a process which produces an acute emphasis on personal narratives that are appropriated by and for the new story of the nation.[21] In this process, new meaning is ascribed to the past, producing a new narrative of it, a new history, a new public record, forged out of the

---

media politics. Of course this always means that a media image may be leached of its specificity and become generic, the symbol of a larger (and not always totally applicable) social issue or narrative.

17 Investigative journalists also took a dominant role in revealing the atrocities of apartheid in published accounts, which often detailed information which they had gathered under apartheid but been unable to publish. See for example the work of Jacques Pauw (1997) and Keane (1994). Pauw also worked for the SABC's, *Truth Commission Special Report*.

18 In this edition on 12th May, 1996, du Preez introduced the case by saying, "The murder we investigate and solve tonight happened 15 years ago. Nobody has been charged with it." The Mxenge family were among those who opposed the TRC process, demanding legal justice instead, but they did take part in this program.

19 These kinds of staged meetings appear in other media representations too. For example, see Edelstein's photo-essay where she posed Mrs Mananki Seipei, the mother of murdered Stompie Seipei, with Jerry Richardson, who as a member of the Mandela United Football Club, was convicted of the murder (1999: 113). In the text, Edelstein says that this was at the request of Richardson himself, and goes on, "I felt a strange combination of emotions as I retreated under my black hood to check the inverted image on the glass" (ibid: 111).

20 On this issue of forgiveness, see for example the *Mail and Guardian* (1.11.96) story of the wife of Harold Sefolo talking about this as a limitation of the TRC.

21 See also Landsman and Ciborski 1992; Boyarin 1994; Healy 1997; Goodall 1997.

confrontation between the passing official history and alternative histories (Clifford 1997).[22] Opposed to the exclusions of the founding myths of the old South African nation state, this history provided a new basis for national imagining but also legitimated it in symbolising and enacting the postcolonial state's claims for an inclusive nation. Here the TRC, especially in its media form, was also fixing just what *was* relevant history for the imagining of nation (cf Papadakis 1998: 155). The apartheid past is the relevant history for the new South Africa, just as settlement and conquest history was history for various past colonial forms of nation.[23]

This new history asserted that South Africa under apartheid had been a place of systematic and pervasive abuse of human beings, abuse which was consistently denied and concealed by the state and its agents, abuse which extended beyond its borders and which ultimately produced counter-violence until stable rule was no longer viable. The liberation struggle itself becomes an acceptable dimension of history in such a representation of the past, though it too involved abuses and violations. With this narrative in place, no longer was it going to be possible to *deny* other experiences under apartheid (cf Hamber 1997a).[24] A dominant media take on this new version of the past was that it could provide unity on which to rebuild nation. A good example of how this was conceptually framed came in the *Mail and Guardian*'s article on the final report (5.3.99) which argued that, "It is a 'people's report' in which the nation has spoken to the nation out of its suffering, presenting this country with the glorious opportunity to strike for unity across the racial divides of the past."

The new history which emerged was clearly still linear and objectivist, inserting new information, interpretation and experience into the gaps of the old official history, at the same time replacing those elements which had been demonstrably 'disproved'. Dominated by investigative journalism, the media TRC's focus was on revelation and confession to fill gaps in a story (with a beginning, middle and end), rather than developing alternative forms of self-expression and discourse (cf Juhasz 1995; Huesca 1995). Ultimately it aimed to provide a more *complete* record through the inclusion of accounts of events which had either never been documented or were suppressed and the perspectives of people previously silenced by the collusion of police, mortuary workers, judges and medical workers with the very system which produced the violations. Although it sought to be inclusive, as was required for it to become a shared history for the new nation, it is nonetheless objectivist history (Landsman and Ciborski 1992), replacing one narrative version of the past with another.

The final report makes this objectivist stance clear, which is exactly why it attributed blame to 'both sides', incurring the criticism of the ANC in particular. It is a version of the past which does not fully capture the realities of apartheid for many people, as we shall see in

---

[22] Krog's Afrikaner formulation of this is the need for half memory to be made whole (Boraine and Levy 1995: 115).

[23] Asmal et al (1996: 8) explain this relevance, and indeed the TRC investigation beginning in 1960 with Sharpeville, by arguing that it was here that South African racist rule as apartheid became bad in its own unique way in global terms. They go on to argue that, even the first Nationalist government in 1948 can be seen in continuity with the colonialism and racist exploitation which dominated "developing countries" throughout the world. In the 1960s however, there is a qualitative departure in South Africa, with the criminalisation of opposition, "brutally slaughtered, as at Sharpeville, rather than heard." Sparks (1990: 235) also points out that it was after Sharpeville's brutal repression that black nationalist politics and the struggle against decades of oppression moved from non-violence and the civil rights campaign form to armed struggle and "a civil war of sorts". These qualitative shifts then begin the historical period which concerns the TRC, and which has become the past against which the new South Africa is to be imagined and defined.

the next chapters. It is, as Scheper-Hughes suggests, an agreed "good enough truth", which she describes as "a narrative that will at least place black and white South Africans, Afrikaners and English-speakers...on the same map rather than living in different nations across the road from each other" (1998: 10).[25] This "good enough truth" is exactly the basis of what I am calling a shared history, a representation of the past which can at least relate together the multiple histories and experiences of people in South Africa.[26]

Emerging in the media TRC then is a broad, generic official history, which, as Koch argued in the *Mail and Guardian* (24.12.96 - 9.1.97: 10), shaped the way people could now describe their past by providing a version of that past, a representation which even those who had claimed not to know could agree "was the truth". This version of the past does not owe its content *directly* to the ANC or other elites, despite NNP and IFP charges to the contrary. There *are* strong tones of the ANC narrative of violence incurred in the liberation struggle against oppression, but this could easily result from the *form* of the hearings and the need to establish particular contexts and motives for gross human rights abuses.[27] As Bozzoli notes, there was a very real tendency to couch township violence in particular in liberation struggle terms (1998), although the particularities of the local histories itself were often lost in the media event.

Given the huge problems of overcoming the silences and the misrepresentations of the past in order to reach this new history, a transformative conceptual mechanism was necessary within the whole narrative process, to alter people's perceptions, especially those people who had unproblematically accepted the apartheid regime's account of events. This mechanism was the production of the uncanny, and the next section will argue that media had a direct role in its generation and deployment.

Gelder and Jacobs (1998) take the concept of the uncanny from Freud in order to explore the ways in which the Aboriginal sacred in Australia has the power to render the familiar unfamiliar. Constituted in this way the uncanny has the power to transform and to generate reflexivity (1998: Chapter 3). Even within its limited frame of human rights violations, the TRC had exactly this potential to render apartheid, the state's practices under apartheid and

---

[24] This inability to deny (although not without some further repression) was hailed by many in the media as the signal success of the TRC. See for example "A bargain with the devil" (*Mail and Guardian* 20.3.97) and "Betraying the 'people's report'" (*Mail and Guardian* 5.3.99).

[25] See also see also interview with du Preez in Australian ABC's *Late Night Live*, 1.8.99. This "good enough truth" is obviously a perennial problem for all elites who take this particular road of constructing shared histories for postcolonial nations. LiPuma (1997: 236) writes of the Solomon Island elites' attempts in this regard, that they are perfectly aware of the problems of the imperfect version of history which resulted, but feel they have no choice within the wider aim of nation building.

[26] There is I think, support for this notion that the TRC provides an inclusive account of history (or a part of history) which can be owned and shared by all South Africans, in the fact and content of the protests against the privatisation of the final TRC Report, which were posted on the TRC website. These were prompted by the removal of the Report from the TRC website when publishing rights were granted to Juta and Company.

[27] It is probably also directly related to the TRC's focus on political violence. Either constructing oneself as an 'innocent' victim of apartheid violence or as a perpetrator of violence in the name of revolutionary aims was consonant with the identification of political violence as legitimate. At the same time, there are consonant patterns discernible in the other side of testimonies, where victims (like the sisters of one of the Magoo's Bar victims) stress that they were 'not political' or perpetrators insisted that their violations were politically not personally motivated, which was a condition for amnesty. Jacques Pauw notes at various points in his study of 'apartheid's assassins' (1997), that some perpetrators spoke privately of their activities in quite different ways, compared with the remorse they showed in the amnesty hearings. The political form of the TRC, and its conditions, shaped what could be known and how, since it affected the ways in which people narrated their experiences.

white hegemony uncanny. Nagengast has pointed out that states seldom publicly present themselves as perpetrating violence and terrorism on their own people (1994:120). What they are thus forced to do is represent these acts within "existing, acceptable discourse", using familiar and accepted images of those against whom violence is used. Under apartheid, such images and discourses were typically those of communism and terrorism,[28] both of which were deployed by the apartheid state with more and more vigour throughout the 1980s.[29] What the narratives revealed in the TRC enabled was a deconstruction of these popular apartheid discourses, shown in the very media which had been so crucial in perpetrating them, rendering apartheid itself and the lives of those who had benefitted from it, uncanny. The uncanny was a central mechanism which provided the moral refiguring of old paradigms, which is the very core of shamanising media events like the TRC. At the same time, the media's rendering of the TRC through its own performance genres and conventions was a crucial condition for creating, enacting and elaborating the uncanny. Once again, media provided a crucial dimension of the public TRC, here at the very heart of the Commission's *own* performance of transition and the imagination of the new. The next section explores the media's performance of the TRC as public event in these terms.

Constructing the Uncanny, Brokering a New History

In her account of Americans' reactions to the television image of the end of the Cold War, the fall of the Berlin Wall, Margaret Morse asks, "Is anything more uncanny than to lose a prime foundation of one's identity - the orientation of 'East' and 'West'? In such a situation, an air of unreality or tentativeness hovers over recent events" (1998: 49). This loss of foundation is exactly the terrain of the uncanny in the TRC, and in South Africa, as in America, media representations provided the condition for it. Formative here was media's role in bringing the ritual performance into the homes and thus the mundane worlds of people not able (nor willing) to be at the event itself. Such an act makes familiar and 'known' people who would never be known to people in their normal lives, creating the possibility of seeing them differently. What must be attended to in the analysis of the uncanny in the media TRC though is how this making known and familiar is very much a technical effect by media. Who becomes known to the public, how they are known and what roles people come to take in the new story of the nation also depend very much on the processes and practices of selection and representation used in media production. The public representation of both human rights violations and amnesty hearings were involved, though in different ways.

From its very beginning, the TRC's narration of the nation occupied the terrain of the uncanny. Its whole purpose of narrating the past automatically involved challenging the

[28] Theissen and Hamber (1998) show how it was exactly the discourse of terrorism which produced enormously high white approval for government and security actions in the 1970s and 1980s. They cite a survey in 1984, for example, in which 90% of white South Africans "felt that the government's combating of 'terrorism' was good or very good" (ibid: 2).

[29] It is interesting that in the 1999 elections, the New National Party was still using the threat of communism to persuade voters; surely evidence of the charge by several journalists that the Party did not understand the new South Africa.

myths (Nagengast 1994: 120, 121)[30] of apartheid, the familiar and taken for granted understandings about the world on which the regime depended. At the same time, its *media* arena came to significantly affect the uncanny as it emerged to have effect in the public domain. Through its own genres and conventions, the media's performance of the TRC multiplied and deepened the uncanny effect of the hearings. Particularly formative in the media TRC's construction of the uncanny were news and current affairs practices and conventions. Just *how* the media TRC constructed the uncanny was influenced and shaped by selection processes, interpretive strategies and narrative forms and themes derived from routine news and current affairs practice. This construction and deepening of the uncanny through media techniques was a fundamental dimension of the brokerage function of media personnel and requires further exploration.

Earlier we saw how the TRC was eminently newsworthy in journalists' professional terms because it fulfilled many of the primary news values which journalists hold. The application of news values also affected the content of the media TRC, particularly its scope and range. Here one is reminded of Felman's exploration of narrative and violence in Northern Ireland, in which he says, "The event is not what happens. The event is that which can be narrated" (quoted in Malkki 1995: 107). Malkki uses this to understand how some happenings in the refugee camps "melted", while others persisted in narration to become events. In media, particularly news media, it is the ability to be narrated in terms of news discourse and news values that determines which happenings will become news events and which will therefore never emerge as integral elements of the media TRC. This principle gave greater media worthiness to some events and stories than others, clearly no less disturbing to the individuals, no less an integral part of apartheid, or the TRC. It was those cases that in some way fitted accepted news values or could be represented best in terms of journalistic convention and performance that routinely dominated, to become the TRC for the media public. In this way the number of cases which the TRC had already reduced was further reduced for the media TRC. The repetition of some very high profile stories, which is also an accepted aspect of news practice, further produced the complaint that many of the cases before the TRC were already well known. This was aimed more at the media TRC than the hearings, which did hear many people unknown outside their communities. As stories that emerged time and again at various points of the media TRC, these stories became very well known, their details familiar and, through media, part of people's everyday lives.[31]

In this media redundancy of particular stories we can see the operation of routine news practices and interpretations, the work of news personnel as cultural brokers. Newsworthiness tends to adhere to events, so that those that had already had high media exposure were again more likely to be covered. Sometimes too, the victim made the story newsworthy, as in the case of the township killing of Fulbright student Amy Bliel or politician Chris Hani and of course Winnie Madizikela-Mandela was in a league of her own in newsworthiness. At the same time, all of these stories also fulfilled news media's tendency to select those events that

[30] Nagengast is here using myth in the sense used by Barthes (1963), rather than the popular meaning of untruth. This is also compatible with Levi-Strauss' anthropological use of myths as stories that seek to address and resolve deep contradictions in society (1978).

[31] The testimony of the barmaid of Magoo's Bar was played over and over in media forum after media forum. She was an excellent media figure at the witness table, very articulate, very self possessed and with a very dramatic style of evidence presentation with which she graphically described the events and the scene; an obvious choice for seasoned news and current affairs professionals.

could be narrated in their terms and through their conventions of news making and telling. This is the heart of the terrain on which the uncanny was produced in the media TRC, because it was some of these conventions of news which operated as "a humanistic, universalizing representational practice" (Malkki 1995: 11).[32] Many analysts[33] have pointed out how many of the conventions of news narrativisation are defined and used by media in ways that universalise, appeal to *sameness* rather than difference. As elements of a universalising representational practice, they dissolve boundaries and differences, or render them secondary to the "universally human" (Malkki 1995: 10), at least that which is perceived to be so. This is exactly a condition on which the uncanny in the media TRC rested, since it made possible seeing beyond the myths of difference and taken for granted assumptions of apartheid. The rest of the chapter explores how the media TRC can be understood to be constructed through such a representational practice, identifying which conventions of news making were most powerful in this regard.

One of the routine feature of news is constructing news events through narrative.[34] Bird and Dardenne (1988) distinguish between the chronicle and the narrative form of news and argue that the narrativisation of news events is primarily a moral rendering (see also Jacobs 1996: 381). What news narration provided for the media TRC was the particular Manichean frame which is usually deployed to achieve this moral rendering (Ward 1995). Emerging in both kinds of hearings, human rights violations and amnesty, this frame constructed 'sides', even though at many times just what the terms 'good' and 'evil' meant was in dispute (How could men who thought they were acting for the good of their country be guilty? Can a person be doing evil if they were just following orders?). Thus the routine practice of framing news meant that torture and barbaric forms of murder and assassination were elaborated in dichotomous terms to provide a *moral* frame. The construction of 'sides' that resulted was now in moral, human terms rather than the political alignments of apartheid racialised ideology, a crucial reinterpretation of the ways in which good and evil *had* been attributed to people and practices under apartheid.

News narration does not just frame the news, it also brokers it through grand themes and archetypal narratives as a way of understanding events.[35] In the media performance of the TRC, themes like the individual struggling against adversity[36] or heroes fighting villains inflected news stories in ways which produced the uncanny. Thus Gideon Niewoudt, the

---

32 Malkki is particularly concerned with the photographic representation of refugees, which she argues constantly reproduces the figure of the woman with her child to construct and evaluate displacement in ways which universalise, despite "displaying differences and 'cultural diversity' among refugees" (1995: 10) at the same time. She suggests that more work needs to be done on this form of representational practice, which is what, in another context, I have sought to do here.

33 Hall et al 1978; Gruneau and Hackett 1990; cf Bourdieu 1996.

34 Jacobs 1996; Bird and Dardenne 1988; Hall 1984

35 See Jacobs 1996: 383; Bird and Dardenne 1988; Hall 1984; see also media event analysis, for example Dayan and Katz 1992; Handelman 1990; Stam 1992.

36 Struggling against adversity here could again extend into the TRC process itself. A very good example was the outrage expressed in the press at the further silencing of victims by the court injunction won by policeman Gideon Nieuwoudt. Jacques Pauw's account of the long delayed testimony of the mother of one of the high profile victims, Sipiwo against Nieuwoudt conveys exactly this media theme of triumph over adversity:

> A few weeks later, Joyce Mtimkulu, flanked and supported by her husband and family members, took the stand at the TRC. She swore an oath, and told the tale of her son who had been broken by those who professed to maintain law and order and uphold Christian values. She read passages from his diary, took the contents from the plastic shopping bag and pointed to the pieces of flesh still attached to his hair. When she was finished, she cried again, but this time out of sheer relief. A nation listened and embraced her with understanding and empathy (1997: 27).

policeman from the Eastern Cape implicated in many abductions and killings who was once a hero of apartheid, was now clearly constructed as villain (one story focused on how he used to dress as a priest to visit the families of those he had killed)[37] and Harold Sefolo as a hero for his bravery as he went to his death singing "Nkosi", then the ANC hymn, now the national anthem.[38] The dominant narrative theme of the quest, after truth or justice (however defined) was another example. Although these themes were clearly and powerfully constructed and narrated in ways that entailed specifically South African discourses, the liberation struggle, township politics, white legitimation and even the TRC itself, they also elaborated what are seen as universal truths, things like love, man's inhumanity to man, life and death, right and wrong. The embedding of grand themes in news stories thus can also be understood in terms of Malkki's humanistic universalising representational practice and a strategy for the generation of the uncanny. Such themes also relate to another, central, dimension of this practice to emerge in the media TRC.

I argued earlier that the TRC provided the perfect form of media event to be constructed and performed through news conventions. There is another way in which this is also clearly the case. All news media are very poor at handling structure and system; their temporality, preferences and constant demand for the new mean that they will always choose event and personality over structure in their accounts of the world. In the TRC though, media were presented with apartheid, a structure, but apartheid told through individual narratives, the personalisation of a system (cf Hamber 1998b: 2). Personalisation in the media TRC lies at the heart of the terrain of the uncanny, here achieved through the practices and strategies of news making. Here one would want particularly to note the power of the visual image; the significance of video footage and photographs which provided the 'human face' of the news stories, affirming their humanity and reality. Through attaching image to narrative, remote news figures were turned into *persons* who could be 'known' and empathised with by those watching, listening and reading in their homes; mothers weeping for their lost children, the good high school student ambushed and murdered by apartheid security forces, the ANC activist who was also fine family man who cared deeply about his home and children or the church going father who is revealed as a vicious killer and torturer. Cameras captured powerful moments which were often used to symbolise the whole, often precisely those which could be interpreted in terms of the grand universal and human themes elaborated above; grief, love, despair, remorse. Particularly noteworthy here in terms of the uncanny was the revelatory power of the image in one of the amnesty hearings in which police agent Jeff Benzien demonstrated the 'wet bag torture' on one of his former victims. The shocking image of a black man lying on the floor, his hands behind his back, with a huge man sitting on him gradually pulling a bag over his head and tightening it (until he passed out) has been described by many as the "seminal moment" of the TRC, the shocking moment of realisation about the past and the jolting of any complaisance. Graeme Simpson of The Centre for the Study of Violence in South Africa, argues that this was the "deep reality check" (*Lateline*, September 1998), even for those white South Africans who had remained in denial during much of the process (whom he compared to Nazi sympathisers immediately after World War Two).

[37] See "The 'priest' from hell" (*Mail and Guardian* 1.5.99).
[38] See "ANC man who went to his death singing" (*Mail and Guardian* 1.11.96).

The Benzien hearing is an example of one of two vital personalisation strategies in news making, which were highly significant in producing the uncanny in the media TRC through their contribution to a universalising representational practice; the use of drama and of emotion. In many ways these are secondary effects of the convention of rendering the news *as* narrative. Despite its categorisation as an information genre, news has always found a place for the dramatic. For example, events that have dramatic potential in terms of their narrating, such as the Benzien case, are also seen to have newsworthiness in themselves. In a media event of the nature of the TRC, this means that they are more likely to be selected or used often. One example, played on television and appearing in the press of all kinds, was a testimony of a woman lifting her son's shirt to reveal the horrific scars from policemen's bullets, which disembowelled him when he was only 15 years old. Another was a woman who was captured on camera revealing the massive wounds to her leg, received when she was shot on the way home from church. At a different level, a popular media story was the dramatic case of a man blinded under torture (and who continued to be harassed even after this) saying that he felt as if he could see again because he had been able to reveal what had happened to him.

In the domain of the uncanny, it was the building of drama *into* the narrative that was most significant. The moments of the media TRC were often rendered in terms of drama, by inflecting news through its principles. A simple example is the very language used which dramatised the stories in ways that profoundly affected their contribution to the greater narrative of the nation being constructed through the media TRC. Thus evidence is often "startling" (even where it was quite well known), the TRC as a "theatre of pain and catharsis" (*Mail and Guardian* 19-25, 4.97: 5), and hearings are described in terms of "shattered lives" (*The Star and SATimes* 19.3.97: 9), "silent anguish" (*Truth Commission Special Report* SABC 2, 21.4.96), "nightmares of the past" (*TC Special Report* 12.5.96) and so on. With such dramatic words events *become* news. From the perspective of audiences, dramatisation demands attention from the audience, inviting a response, whether this is empathy or dislike for the persons emerging as 'characters' in the story of the nation.

Drama as a form of story-telling also contributed directly to the uncanny by rendering structural conflict in terms of persons and universal themes. The narrative themes cited above are the very stuff of drama, which consistently sets up particular situations and relations to explore the grand issues and dilemmas of life and death. More importantly for the uncanny, by dramatising the TRC's stories, the media narrative tended to portray the systemic contradiction and tension of the apartheid regime in terms of the particular, as conflict with protagonists, motivation and personal inflection, in keeping with the form of dramatic interrogation of life and society. The Hani amnesty hearings were a perfect example of the construction of news drama around conflict between protagonists. Here the high profile lawyer, George Bizos was pitted against Gaye Derby-Lewis, accused as an accessory to the Hani murder. Journalist, Angella Johnson describes the confrontation between these two in terms of a contest:

> They faced each other like gladiators in the truth commission arena; two battle-scarred adversaries in a fight to decide if the killers of Communist party leader Chris Hani should be granted amnesty and walk free (*Mail and Guardian* 5.12.97).

This is classic news drama, clearly constructing history out of media conventions of telling news stories in compelling, dramatic ways that invite people into the human dilemmas unfolding on television screens and in newspapers. It almost seems unnecessary for Johnson to later describe the hearing as "pure theatre", but she does.

The second element of news story-telling which powerfully contributed to the personalising of structure and system was the importance of *emotion* in so many of the stories. Headlines often constructed emotion as *the* significant dimension; take that simple "Heart-rending" (*Sowetan* 16.4.96) discussed earlier, or "How They Spoke the Unspeakable" (*Sunday Independent* 21.4.96) and "The Years of horror and a sad widow's plaintive cry for truth" (*Sunday Insight* 4.5.97). Clearly the focus on emotion played out a crucial theme of the TRC itself; the cathartic effect of telling the stories of atrocity and violation in which weeping particularly was seen as a form of closure (Bozzoli 1998). News consistently appeals to emotion at least as much as reason.[39] Max du Preez makes this fit between the TRC and news story around emotion quite plain when he says,

> The Truth Commission hearings were perfect for television journalism. It was not a story about politicians - it was about the way ordinary men, women and children felt about the horrors of apartheid. The TV cameras could take the close-ups of these feelings into every living room in the country (Thloloe 1998: 1; cf Mona 1998: 38).

The close-up of course is the most personalising of all camera shots and here, emotion is less about closure and more about symbolising the TRC itself in some way.

Emotion is also central in the construction of the uncanny. Various scholars have noted its importance in news through its generation of sameness and empathy. In the news emotion is the very *idiom* of the personalising human interest story, which has come to dominate so much news today,[40] a form which is all about appealing to what is the same about people as fellow humans, rather than differences. Emotion was a crucial strategy in the media TRC's humanistic, universalising representational practice which set up the conditions for the uncanny.

Emotion became *both* a dominant theme in the media's performances of the TRC *and* a narrative strategy; a way of telling and part of the story told. There are countless examples from electronic media and the press of people breaking down while giving their testimony, with counsellors handing them paper tissues or simply touching them, holding them. When Tutu first broke down during the hearings, this appeared with photographs on front pages everywhere and also in radio and television news.[41] At other times, the poignance of the evidence of particular people was underlined by scenes of members of the audience wiping their eyes[42] or the testimony being framed by voice over commentary that commissioners had been visibly affected by the evidence. One case which had all of these was that of Mrs Timol asking the Commission to find out what really happened to her son, Ahmad, who died in

---

39 See Street 1998; Altheide and Snow 1991. Altheide and Snow argue that this appeal to the emotions rather than reason is exactly a feature of media politics. See also Postman (1985).

40 See Bourdieu 1996: 51; Becker 1995; Langer 1992. Bourdieu argues that this is a factor of competition in the market. Human interest stories appeal to the widest possible market, so these come to be preferred over hard news which has a much narrower appeal.

41 These images were used again later when Tutu's illness was revealed.

42 This was also worked into many press stories. See, for example "Young priest 'icon of forgiveness'" (*Weekend Argus* 15.6.96:9).

police custody. The English news (SABC3, 30.4.96) framed the story verbally over Mrs Timol crying at the witness table with her son's arm around her, thus "The Commissioners were moved to tears listening to frail Mrs Timol as she relived the day that her son died in police custody." Here too, as in so many other instances, we see the representation of the woman and her (lost) child to construct and evaluate the apartheid past in ways which universalise (cf Malkki 1995: 10).

Emotion remained a news theme throughout, emerging also in the amnesty hearings. A good example was "Tears From the Tough" (*Mail and Guardian*, 8, 26.2.97:9), in which a Vlakplaas policeman cried during his evidence.[43] This story combined both elements of emotion in the construction of the media TRC; emotion dominates the telling of the story and is its point. The article specifically argues that this is what the TRC is about, transforming people, so that emotion is the sign of repentance, transformation and thus redemption.[44] It is a sign of the policeman's humanity. A further sign that the TRC was capable of rendering transformation in the country, uniting on human grounds, came through the accounts, particularly in the first year, that journalists too were moved to tears by what they heard and saw.[45] This, of course, takes part of its meaning from journalists' professional reputations as having seen everything, being immune to horror and misery, a view that Krog says was part of South African journalists' own self images *before* the TRC (1998: Chapter 2).[46]

Emotion, breaking down, thus often *became* the story, precisely because this provided evidence that the TRC was working. Often it was this focus on emotion which united narratives which were quite separate in the hearings themselves to run together in the media account.[47] In such a strategy, emotion is universalised. Its deployment constructs sameness *across* divisions in society, appealing to a unity on the basis of humanness, or at least asserts that it is what is the same, feelings, that is important, pushing difference into second place. This appeal to unity is a strategy in media which directly aligns it with the aims and strategies of the TRC itself.[48]

---

43 The very first amnesty hearings in May 1996 were also reported as an example of reconciliation happening already, partly because the applicants had asked for forgiveness but perhaps most powerfully because one of them was "tearful", something that is visible in the accompanying photograph of the testifying policeman (for example, "Picking up the Pieces", *New Nation*, 24.5.96:5).

44 These cases of repentance were perhaps more powerful precisely because amnesty applicants did not have to ask for forgiveness, only to demonstrate that their motives were political.

45 For example "Overwhelming trauma of the truth", *Mail and Guardian* 24.12.96- 9.1.97; Krog 1998; "I have never seen my husband's grave, *Weekend Argus* 20-22.4.96.

46 The emergence of emotion as *the* central theme in a media event was also clear in what Walter et al (1999) call "the Diana event", the mediated death and mourning for Princess Diana in 1996. Here, the emotions of the public, most of whom had never known her in life, became a major narrative theme. Kitzinger notes that, in reporting the event, journalists and news presenters made their own emotions part of the story, again a feature of the media event that took its power from their ideas of professionalism.

> The emotional tempo was raised by the way in which journalists and newscasters expressed their own shock. Media professionals traditionally scorn emotional displays. However, in this case several media professionals publicly stated that they were close to tears. Mingling with the crowds of mourners and even working with the raw footage itself seemed to disturb those in front of and behind the screen (1999: 72).

47 See for example "Scenes of tears and unbelievable courage ", *The Star* 58, (16, 1997:2) in which emotion unites the stories of Amy Bliel's parents and Dawie Ackermann whose wife died in the St James Church massacre. The two stories run almost seamlessly together because, with the details of what happened removed in this media account, they become stories of tears, forgiveness and courage in adversity, perfect human interest stories.

48 At the same time there was resistance to the TRC in some quarters, precisely *because* of the primacy of emotion in the media TRC and the way in which it rendered people testifying. As one Cape Town activist asked, having said that people already know because they've already been told, "Why must we now come again and sit in front of television cameras so everyone can see us cry? I'm not a victim, I'm a survivor." ("Cape activists resist truth

The personalisation of news through narrative strategies of grand themes, drama and emotion clearly constructs the media TRC within Malkki's humanistic, universalising representational practice. As dimensions of such a practice, they dissolve boundaries and differences, or render them secondary to that which is seen as universally human. But where both Hall et al[49] and Bourdieu (1996) see such a phenomenon in terms of the *leaching* of politics, in the media TRC it is reframed in terms of a postcolonial politics to be political in a very significant way. This practice of universalising through a range of news conventions is central in the media TRC for its construction of the uncanny, through which are achieved the political and moral aims of the TRC as public event. The assertion of a deeper unity for the nation, one that *subsumes* difference, is directly located on the terrain of the uncanny which the media TRC's universalising practice constructs. How is this so?

Operating within the pedagogic and moral frame of the media TRC, this practice and the strategies which constituted it rendered apartheid's Other *human*, not different but the *same*; people like you and me, with attachments and feelings the same, living in pain and sorrow, grieving, lost (cf LiPuma 1997: 225). They weren't all communists or terrorists and therefore threatening, just by virtue of race and culture. They *were* treated with cruelty and horror, in ways that the state systematically denied were happening. In turn, this also rendered *home*, South Africans' being in the world, uncanny, at once still familiar but now totally unfamiliar.

> An uncanny experience may occur when one's home is rendered, somehow and in some sense, unfamiliar; one has the experience, in other words of being in place and 'out of place', simultaneously (Gelder and Jacobs 1998: 23).

In all of its dimensions, but especially its strategies of personalisation, the media TRC provided the conditions in which this uncanny experience could occur, especially for white South Africans. It deconstructed the myths and revealed other truths.[50] It forced, or attempted to force, recognition (LiPuma 1997: 225). In its telling of the TRC stories, 'terrorists' became heroes or ordinary people persecuted and harmed by the regime. Radical difference is rendered as human sameness, a construction of the human subject as the new national subject that incidentally, also directly supports the political form of nation.

Also part of the uncanny is the other side to the appeal to universal humanness within a moral frame provided in the amnesty hearings. Here was revealed that men who were publicly known as respectable, caring family men had secret lives as torturers and murderers. Scheper-Hughes (1998: 13) has argued that, the "TRC has provided the world with unforgettable images of culture inverted and a world turned upside down". Authorities and powerful figures were revealed, through media's rendering of witnessing, as abusers, inverting the system's own self definition of them as upholders of law and order (Supriya 1996: 100). Thus in the media reports we routinely find appearance and reality contrasted, with it being said of the infamous Craig Williamson that he had the "appearance of a

hearings", *Mail and Guardian* 26.4.96). I heard this in the District Six Museum in Cape Town too, in various discussions about the TRC hearings as seen on television.

[49] Hall and his co-authors (1978: Chapter 3) for example see the practice as obscuring the workings of social system, inequality in representation and social situatedness in news accounts of crime.

[50] Of course, media revelation can't automatically secure the overturning of myths inculcated under apartheid. As Matthew Phosa said of the revelation that police were linked to the Joubert Park bombing which PW Botha blamed at the time on the ANC, lots of people "will still remember the ANC was to blame" ("Police linked to 1987 bombing", *Mail and Guardian* 5.7.96).

trencherman rather than a policeman into personal betrayal" (*Mail and Guardian* 11.9.98). Another policeman, Hennie Gerber, was described as looking "more like a casually dressed insurance salesman than a convicted murderer. He seems a nice enough guy, the kind who likes to braai and drink a couple of beers" (*Mail and Guardian* 20.3.97).[51] These formulations seem to quite deliberately render the uncanny, as they transform the familiar, known, church going, respectable man into the unfamiliar, the unspeakable, the unthinkable, in front of the public.[52] Jeff Benzien even told the TRC, faithfully reported on camera, that he could not recognise himself. "What type of person am I?" is the question he asks himself, he said. This inability to name or recognise oneself is surely a radical example of the uncanny and one that encapsulates its power as a mechanism to enable the reimagining of nation.

The media uncanny contributes to the media TRC's narration of nation in several ways. It destabilises accepted understandings of the past system by rendering official accounts of it strange and turn accepted constitutions of identity under apartheid on their head. It is thus the condition for the emergence of a new account, a new inclusive history that at least seeks to place all South Africans within the same narrative frame. In this register, it supports the moral process of the media event, signalling transformation and the emergence of a new basis for the nation. The emergence into the media spotlight of significant apartheid assassins (Pauw 1997), revealing in media styled narrative just what they did and at times their sorrow, is made to demonstrate that such behaviour is no longer acceptable. This is precisely how figures like Benzien and Williamson become such important figures in the media event. Exemplary figures in news terms,[53] in the media event they enact in testimonial performance the passing of the old and the emergence of the new, as a totally transformed basis for nation.[54]

As Gelder and Jacobs (1998: 23) argue, the uncanny is not a dichotomy. Rather the familiar and the unfamiliar, the home and the unhomely "inhabit each other", so that while the uncanny is reflexive and destabilising, there is no guaranteed outcome. The writers unpack the complex and contradictory workings of the sacred uncanny in Australian political discourse today. In South Africa we can see how the media TRC's terrain of the uncanny is occupied by a range of people and outcomes. For example, we can see it operating in both those who hardened their position against the TRC and those whites who argued they did not know what was happening under apartheid. The latter too became part of the media TRC, from the letter by an Afrikaner to the TRC asking for forgiveness for turning a blind eye to what was happening[55] to those who claimed they really did not know. The latter would appear to be arguing that they really had believed the state's representation of itself, so are reacting to the media TRC's deconstruction of the myths about communism and terrorism which

---

[51] Even the *braai*, the barbeque associated with the light hearted social occasion with family and friends was rendered uncanny in the often used media story of the Vlakplaas officers enjoying a braai next to the burning body which they had set alight to destroy the evidence.

[52] See Scheper-Hughes 1998: 10; Krog 1998: Chapter 6. This whole topic and how families were crucial to the process of reconciliation as they helped perpetrators to deal with their shame was dealt with in the religious television discussion program *Credo*, 28.4.96.

[53] See "Day of the assassin promises to thrill", *Mail and Guardian* 30.8.96.

[54] To achieve this, it doesn't really matter whether they are repentant. Even where news stories focus on the protests of apartheid foot soldiers, that they are being blamed for 'just following orders' or that they are 'badgered' by the TRC (for example *Sunday Independent* 1.12.96: 5), the signalling of a new basis for nation is still achieved in media.

[55] This letter was read out by Tutu during the TRC proceedings. It thus became part of the radio presentation but also was featured in news bulletins and services.

sustained and legitimated the state under apartheid.[56] On the other hand, in this area, there are now many, like one of the sisters of a women killed by the Magoo's Bar bombing, who claim that they had never supported apartheid (see *Truth Commission Special Report*, SABC2, 12.5.96). At the same time, those in the NNP who used the media TRC to construct themselves as a now disempowered minority being witch-hunted,[57] were also clearly reacting to the construction of the uncanny.[58] The uncanny is very complex, never certain in what it secures in terms of transformation. It has facilitated the state's project of generating a new narrative of nation, supplying some at least of the content of newness for post-apartheid South Africa. Less clear is how permanent are its challenges to old ways of thinking about people in terms of ethnicity and race, divisions in society rather than the things which unite.

CONCLUSION

As a shamanising media event, the intention of the TRC was to offer transformation of individuals and of South African society. Its new paradigm for South Africa was based on a moral engagement with the past, which challenged the concealments of the apartheid regime through the testimonies of those whose voices had been silenced for various reasons. Not only did the process suggest a new moral regime for the future, it also produced a new version of the past, a new history, out of those challenges and voicings. Such a history could serve as a shared version of the past as the ground of imagining a new nation, constructed out of its lessons. But the element of sharing involved more than filling in the gaps and silences. It also required altering the very ideologies and maps of meaning (Hall 1980) which had been inculcated under apartheid. Such a transformation required radical mechanisms.

This chapter has demonstrated that perhaps the most radical of these mechanisms was the production of the uncanny in the TRC; the rendering of the familiar unfamiliar and the challenging of the taken for granted identity constructions on which apartheid had depended, particularly among its white constituency. Further, as in other dimensions of the TRC, media played a crucial role in the production of the uncanny for public reflection, the rendering of that which was familiar strange, thus opening up the way for new ways of looking. The

---

[56] This is certainly the interpretation I heard in many conversations with whites, though many will also preface or interrupt their explanations with throwaway comments like, "of course deep down we knew" and then "but we didn't really understand" or "but we were getting on with our lives" (cf Scheper-Hughes 1998).

[57] *The Star and SA Times* 21.5.9: 3; *Cape Times* 3.12.96: 4; de Klerk 1998; Tutu 1999.

[58] Further afield, one can also see the uncanny working in some of the discussion around the elections of 1999. The Electronic *Mail and Guardian*'s chatroom after the election saw a discussion about the election results, in particular the success of the Democratic Party (DP) and the possibility of it allying itself, either with the ANC or the New National Party (NNP). One posting, for example examined the DP slogan "Let's Fight Back":

> To fight back means to fight in order to regain a position of ascendancy you had before - for the DP and most of the right wing South Africans who voted for it, it means a restoration of white Apartheid hegemony. It means fighting back against the positive strides that have been made over the past five years to restore Black people's dignity in this country. It means fighting back against the guarantee of access to a quality education for all South African children. It means a fight back against equity in the distribution of welfare to all South Africans who require it....The fight back is about striking out at the reality of rule by the majority. In apartheid days the DP/PFP did not fight against anything because the Nats were actually doing what Tony Leon so passionately hankers after now...

There are many more postings which make something of the same point, accusing white conservative and liberal South Africans of wanting to regain old privileges and apartheid visions of the world; in my terms of responding to the uncanny by seeking to restore familiarity to 'home' (cf Gelder and Jacobs 1998: 14-15).

performance of the TRC through the conventions, narratives and interpretive forms of routine media practice and performance genres worked to produce sameness rather than difference and so deconstruct both Other and Self under apartheid. Media accessibility also ensured that the reflexive possibilities encoded in the uncanny were experienced as widely as possible.

The uncanny is a condition of the TRC's enactment of transition which I argued earlier, was a necessary feature of the transformation process. At the same time, I have raised the uncertainty and instability of the uncanny, something which was already visible in the TRC as it was in process, in both the public event and its media performance. Further, the uncanny in the TRC (and perhaps in relation to the sacred in Australia also) is often revealed at a price in terms of the new representations which replace or destabilise the old. The construction of the uncanny was both enabling and constraining. In particular, it forced limitations pedagogically, in the kind of history which emerged and its interrogation of the past. The new history too had its gaps and silences, which proved problematic almost from the beginning of the whole process.

The next section moves onto this terrain of problem and contestation in the process of narrating the nation in post-apartheid South Africa. From the story of the nation emerging in the media TRC it shifts to other performances and representational sites which are also seeking to narrate the nation in terms of South Africa's past. The analysis focuses on a range of case studies which seek to address the problems and silences in the media's narrative of nation as it emerged in the media TRC. In these case studies, South Africans interrogate and respond to this media narrative, relating it to other dimensions of the attempts by the state and its elites to build a nation which can be understood as profoundly new.

INDEX

A

B

C

D

G

H

I

J

K

L

M

N

Q

R

S